POLLINATOR PROTECTION

A BEE & PESTICIDE HANDBOOK

Carl A. Johansen & Daniel F. Mayer

Preface

There is an ever-growing need for pollination of the world's critical food supply. Honey, bumble, leafcutting, alkali and other bees play an important and often indispensable role as flower visitors.

Before the arrival of parasitic mites and Africanized honey bees in North America, bee diseases and exposure to pesticides were the most common problems beekeepers experienced as deterrents to peak pollination. While both bee disease and pesticide kills result from outside intrusions into the colony, bee disease is basically a beekeeper problem while pesticide kills involve non-beekeepers.

During his career at Washington State University, Dr. Carl A. Johansen served both beekeeping and farm interests by conducting studies into the effects of pesticides on bees. Dr. Daniel F. Mayer worked with Johansen and continues this ongoing and important work now that Johansen has retired.

This *Handbook* offers a positive set of solutions to a usually negative topic. It advised and even demands cooperation between beekeepers and pesticide applicators. And it contains an enormous amount of information which Johansen and Mayer have collected over the years. Because the bee-pollination conflict knows no geographical boundary, efforts have been made to increase the scope of this handbook beyond the western United States.

In true handbook style, this text has been developed so the reader — beekeeper or applicator — can quickly find the information he or she needs. As such, it should serve as an important step toward protecting the pollinators of our world's food supply.

— *Larry Connor, Ph.D., Publisher*

ACKNOWEDGEMENTS

We would like to acknowledge the many beekeepers, growers and members of the agricultural industry for their help and support over the years. We thank Dr. Larry Connor for his valuable assistance in preparing our manuscript for publication. We thank E. Laurence Atkins, Riverside, California, for suggestions concerning several sections of this book and for many data used in Appendix VI.

We dedicate this endeavor to our wives, Ruthie Johansen and Connie Mayer, and sincerely thank them for all their support.

— Carl A. Johansen and Daniel F. Mayer

4

TABLE OF CONTENTS

15
Reducing Pollinator Damage and Death

16
APPENDIXES

17
Further Reading

18
Glossary

19
INDEX A—CHEMICAL INDEX

20
INDEX B—GENERAL INDEX

Photo Credits

LIST OF FIGURES

INTRODUCTION

Beekeepers are all too familiar with the term **bee poisoning.** However, persons who do not keep bees sometimes confuse the term and wonder if we mean the effects of bee stings on people. For this reason, we have switched to using the term **pollinator protection** as a more positive and clear description of our research subject in recent years. Of course, bee poisoning refers to the effects of pesticides, especially insecticides, on bees. The main exposure for honey bees is for workers foraging in treated crops and mainly involves the chance adherence of chemical residues on the body of the bee. In the case of alfalfa leafcutting bees and alkali bees, it is the egg-laying females which are primarily exposed to pesticide residues.

If the entire field force of foraging worker honey bees is killed by an insecticide, the beekeeper may lose a super of honey (50-60 lbs). The fruit or seed grower may lose a week of good pollination activity in his orchards or fields. But, often the bees can recover because of the reproductive capabilities of the unexposed queen.

However, when insecticide-contaminated pollen is carried back to the hives, much more severe losses can

bee poisoning

pollinator protection

pollen
contamination

occur. We call this the **contaminated pollen-queen supersedure syndrome**. It can lead to total destruction of the colony.

It's impossible to determine the loss in terms of food production, financial value of the bees, or the discouragement of future beekeepers from bee poisoning. However, the annual cost in the USA of losses due to honey bee poisoning and reduced pollination was estimated at $135 million in 1980.

The reason for this book is education. Beekeepers know pesticide damage to bees is their number one problem in many areas. All surveys during the past 35 years show this very clearly. Even lesser concerns such as winter kill and failing queens are often related to bee poisoning. We think informing beekeepers, growers, and applicators about bee poisoning and how it can be reduced or avoided is the best way to attack the problem.

I.2 — Extreme economic losses result when pollinating insects are killed. Here the beekeeper lost valuable colonies, and the grower lost essential pollinators.

HISTORY AND ECONOMICS OF BEE POISONING

The first recognition of a pesticide problem for bees began in the early 1870's when an unusual malady appeared among colonies of honey bees. During spring, dead bees piled up around the hives, many colonies continued to weaken into the summer, and often they died out completely. The unidentified problem and the use of paris green to control the codling moth on apples and pears appeared on the scene at the same time. G. M. Thompson stated in 1881, "I applied paris green to a pear tree in bloom and killed many bees."

paris green

A. J. Cook in Michigan and F. M. Webster in Ohio were the first to conclusively prove that honey bees were killed by arsenical insecticides applied to fruit trees in bloom during 1889 to 1896.

About 1920 two new factors arose: (1) the development of cheaper calcium arsenate dust, and (2) the use of the airplane to apply insecticides. The boll weevil had become established in the South, but repeated applications of calcium arsenate reduced its damage. The land of cotton covered more territory than the orchard areas and calcium arsenate dust was more hazardous to bees than the arsenical sprays used on tree fruits. Consequently, honey bee losses became much more extensive.

calcium arsenate

Another major factor was reported by N. E. McIndoo and G. S. Demuth in the United States in 1926 and by F. A. Herman and W. H. Bittain in Canada in 1933. These investigators showed that insecticide-contaminated pollen was the principal source of damage to bees and brood: the major cause of continuing debilitation and total destruction of colonies. Since 1946, the new synthetic organic insecticides have brought new and varied

DDT problems for the beekeeper. DDT, especially when used in the emulsifiable concentrate formulation, was not a serious problem; but after its loss because of environmental contamination, many DDT replacements have caused severe damage.

The first organochlorine to cause losses of thousands

carbaryl of colonies was dieldrin in the 1950's. However, carbaryl (Sevin) was the first of the newer organic insecticides to cause severe debilitation of colonies similar to the effects of calcium arsenate. Pollen contaminated with the dust formulation of carbaryl continued to kill bees from one

1.1 season to the next when it was stored in the combs.

Young nurse bees are often killed by pollen which is contaminated

Parathion began killing bees in the late 1940's. However, it and other organophosphorus materials such as malathion and diazinon, often killed the field force without contaminating the hive and causing complete

colony destruction. It remained for a micro-encapsulated formulation of methyl parathion (Penncap-M) to cause long-term hive contamination problems in 1975. The change from the usual EC spray form of methyl parathion to the micro-encapsulated formulation resulted in the loss of several thousand colonies the first year in Idaho, Michigan, New York, and Washington. Like Sevin (carbaryl) dust, it proved to retain a severe hazard from one season to the next.

micro-encapsulation

Currently, the new pyrethroids are generally broad spectrum insecticides. However, a few are relatively safe to bees, and some are repellent enough to bees to provide safety. The dramatically reduced dosages of the 2nd and 3rd generation materials appear to promise a better day for beekeepers in the future.

pyrethroids

Repellents have been studied since the early 1940's with the idea of using them to safeguard bees from pesticides. Recent developments have shown the bee repellent effects of insecticides themselves to be a more important consideration. In some cases the pesticide may only cause a low to moderate kill of bees, but may severely reduce the yield of the treated crop because of interrupted pollination.

Development of insecticides and/or formulations with greater favorable selectivity for bees appears to be the best long-term answer to bee poisoning.

Economics of bee poisoning

Cotton is king and available evidence shows it's the most dangerous crop for bees around the world. Reports from many countries on the killing of bees by pesticide applications to cotton abound. Cotton once was the main honey crop in Egypt, but the frequent spraying of cotton now means that beekeepers dare not leave their hives near cotton. In 1942 in Arizona, 7,231 colonies were lost from pesticide applications to cotton. In 1967, Sevin (carbaryl) applied to cotton killed an estimated 70,000 colonies of bees in California, about 15% of all the colonies in the state. Bee kills have occurred on just about every crop that blooms and from many flowering weeds. Here are some examples:

colony losses due to insecticides

• Washington, 33,000 colonies were killed by Sevin (carbaryl) on corn in 1967.

• In Arizona, the number of honey bee colonies dropped from 116,000 to 60,000 between 1963 and 1977 due to bee kills.

• In 1967, the estimated national loss from all pesticide poisoning was 500,000 colonies. Between 1976-78, 69% of Washington's, 56% of Arizona's, 16% of California's, and 6% of Wisconsin's honey bee colonies were lost or damaged by pesticides.

• In 1984, 24 bee poisonings were reported to the Washington State Department of Agriculture. This represented a loss of over $1 million to the beekeeper.

A comprehensive 1967 economic survey of 30 commercial Washington beekeepers showed the economic effects of insecticide damage. The operators suffered a $204,000 loss from insecticide poisoning in 1967, sustaining a 3.2% loss on investment instead of an expected 11.2% gain they would have accrued without the adverse effects of insecticides. A three-year study in Washington, from 1979 to 1981, showed that 66-79% of all honey bee colonies received at least one kill from pesticides every year. Another well-documented series of heavy bee losses due to pesticide poisoning comes from California, where beekeepers lost an average of 62,500 colonies a year from 1962 to 1973. For example, in 1970, 89,000 out of a total 521,000 were lost from pesticides.

Sevin and Sevin XLR

When Sevin (carbaryl) was first used in Pacific Northwest orchards, one beekeeper lost several thousand colonies in less than a month. As Sevin was registered for use on additional crops starting in the late 1950's, it's impact upon bees became even more devastating. Currently, special new formulations of carbaryl such as Sevin XLR are not nearly as toxic to bees as the older formulations and this has helped alleviate the problem.

impact on other bee species

Other bee species have also been killed. When diazinon was misused for aphid control on alfalfa hay fields while the plants were in partial bloom, many adult alkali bees were killed, causing a 95% reduction in alkali bee larvae in three nearby soil nesting sites. The losses in potential seed production and pollinators totaled $287,000 and the alkali bees had only regained 25% of their initial populations two years later. In 1987, applications of Metasystox-R on blooming alfalfa caused a 90% reduction in alkali bees in four beds, a loss of $500,000. In 1988, alfalfa leafcutting bees in four fields were killed from insecticidal drift. The loss totled $275,000.

Immediate monetary losses caused by bee poisoning are considerable, but the long-term losses in yield of insect-pollinated crops can be even greater. This is particularly true of crops that are dependent upon wild bee pollinators; these may require three years or more to return to original population levels or may be eliminated from a locality indefinitely. Beekeepers often are unable to continue to supply strong colonies of honey bees for pollination service after suffering chronic poisoning losses over a period of years.

1.2 -Near complete loss of honey bee colony follow- ing insecticide exposure.

2   BEES AND THEIR RELATIVES

The **Hymenoptera** are the most beneficial insect group. It contains many of the parasites and some predators which help control insect pests. Of course, it also contains the major pollinators of agricultural crops.

Sawflies are the most primitive kinds of hymenoptera. They have the abdomen broadly joined at the thorax; all others in the group are at least crimp-waisted as in the bees or thread-waisted as in many wasps. Most sawfly larvae are very similar to the caterpillars of moths and butterflies and some are damaging chewing pests of crops and trees.

Ichneumons include many kinds of parasites that lay their eggs in, on, or near the host insect which the larva will ultimately feed upon. Many are of considerable value in the control of insect pests. Typical ichneumons are relatively large and slender-bodied and many of the females have long slender egg-laying organs.

Braconids are usually smaller and more robust in form. Some of these are well-known as aphid destroyers because the females inject their eggs into the bodies of aphids.

Chalcids are mostly quite small and some are minute, less than 1/50th inch in length. Many are

2.1
*Leafcutter bee
(Megachile)*

2.2 -*Leaf-
cutting bee
cutting leaf.*
19

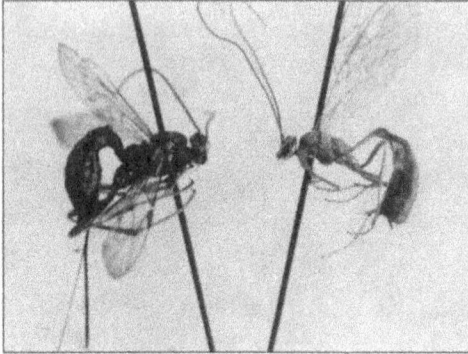

2.3- Ichneumon wasps— Most are parasitic on other insects.

2.4 - A disturbed ant nest.

2.5 - Solitary spider wasp

beneficial parasites and many have been successfully used in biological control programs, especially against scale insects. The woolly aphid parasite, which was introduced into many other areas from New England, does a good job of keeping the woolly apple aphid under control. A few chalcids, such as the alfalfa seed chalcid, are crop pests.

Ants are probably the best known and commonest group of insects in the world. The most distinctive features which separate them from their near relatives are 1-2 humps on the thread waist and elbowed antennae. Ants are highly successful, social insects which have developed some amazing life styles, such as growing fungus gardens and caring for aphid "cows." Many are serious pests of man, but some are effective predators of injurious insects, especially in the forest.

Wasps

Some groups of wasps are **solitary**: they construct a nest of some sort, usually in the ground or a natural cavity, or of mud, and then find suitable prey, sting it, place it in the nest cell, lay an egg on it and seal up the cell. Mud daubers, sand wasps, potter wasps, and spider wasps are familiar examples.

Social wasps, such as yellowjackets and paper wasps, develop colonies somewhat like social bees. Wasps construct paper nests either in the ground or in the open, often attached to trees or houses. Paper wasps are somewhat elongate and slender. Their nests consist of a single, more or less circular, horizontal comb of hexagonal paper cells. Often these are seen attached by a slender stalk under the eves of houses. They tend to be quite beneficial, since they feed their young on chewed up caterpillars or other insect larvae and they are not quick to sting people. Yellowjackets more nearly resemble social bees in their robust form and in producing large colonies in paper nests of several to many horizontal comb-like layers surrounded with a paper envelope. Of the seven common native species in the Pacific Northwest, only 2 are obnoxious pests of man. These tend to shift to scavenging habits in late summer or fall, becoming the number one cause of stinging insect problems. Control methods should be devised to selectively control the pest species, since the others are beneficial predators throughout their cycles. The German yellowjacket (*Vespula germanica*) is becoming the major pest species in many parts of the U.S.

Bees

Bees are functionally separated from wasps by the fact they collect pollen to feed their young. Most have special pollen-collecting devices for this purpose. In addition, many are more robust and hairy than wasps. Bees are

2.6- Yellowjacket wasps are often confused with honey bees

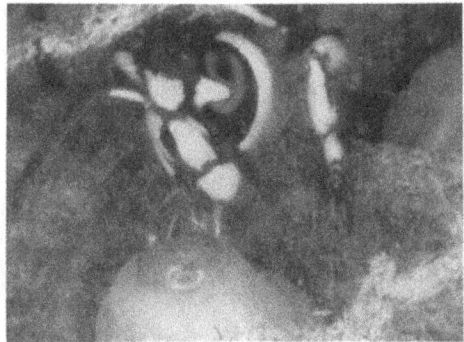

2.7 - Bald-faced hornet feeding a larva inside paper nest

2.8 - Leafcutting bees, showing cutting mouthparts

21

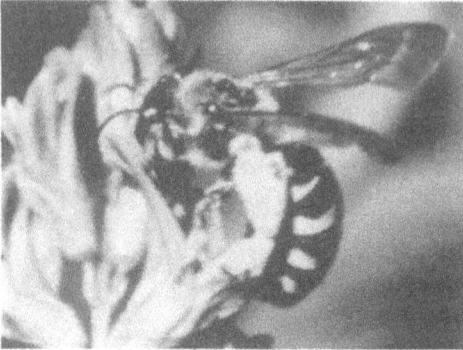

2.9- The alkali bee is one species of mining bee.

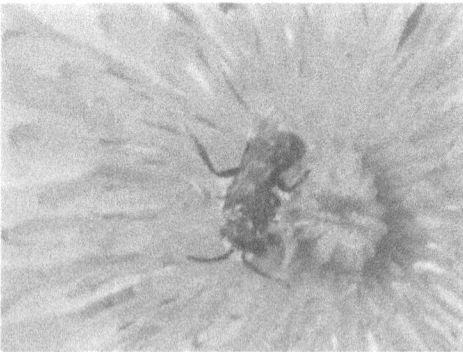

2.10 - Halictus bee

2.11 - Andrena armata, a semi-social bee species.

often divided into two groups: 1) honey bees; and 2) wild bees. Wild bees are solitary or semi-social. Seven common groups of wild bees are solitary and the females produce cells in the ground or in cavities much like the solitary wasps.

Mining bees (Halictidae) are small to medium-sized species which burrow in the ground. Sweat bees often have metallic colors on their bodies. They also prepare nest cells in the ground. Our best-known native species is the alkali bee, an excellent pollinator of alfalfa seed crops.

Leafcutting bees (Megachilidae) make their cells with cut pieces of leaves, either in holes in wood or burrows in the ground. Alfalfa leafcutting bee is the exotic species which is now propagated by alfalfa seed growers throughout the West.

Mason bees, close relatives of the leafcutters, form mud-walled cells in tunnels in wood. Both types pack pollen in a pollen brush of long hairs on the underside of the abdomen. A native species, the orchard mason bee, shows promise as a supplemental pollinator for tree fruits.

Digger bees (Anthophorinae) are usually larger and heavier-bodied than mining bees and many are quite hairy. As their name indicates, they nest in the soil, often with typical mud turrets around the entrance holes.

Carpenter bees (Xylocopinae) are often large and robust, resembling bumble bees, but have dark, shiny abdomens. They are much more prevalent in subtropical and tropical climates where they are important pollinators of tropical fruits. Small carpenter bees construct their cells in the pith of certain plant stalks. Once in a while they become a minor pest of raspberries because of nest construction in the canes.

Cuckoo bees (Anthophorinae) are non-pollen-collecting types which lay their eggs in bee nests, so their larvae develop on the pollen stores of the host. They occur in most bee family groups and often greatly resemble the bees they parasitize.

The main kinds of **social bees** present in North America are bumble bees and honey bees. Stingless bees are near relatives of the Old World honey bees which are mostly found in South and Central America. Only one species has been recorded in southern Texas.

Bumble bees (Bombini) are much more primitive than honey bees. They develop a smaller annual nest of wax cells and only the mated new queens successfully hibernate through the winter. Their annual cycles are quite similar to those of the yellowjackets. We have about 30 species of these large, furry, colorful bees in the Pacific Northwest. Nationally they are particularly important as pollinators of berries and red clover.

The Old World **honey bee** (*Apis mellifera* L. is considered to be the only domesticated insect. Its perennial nest, foraging habits, communication system and complex social organization

2.12- Bumble bee (Bombus)

2.13 - Honey bee worker with pollen on hind leg (corbicula)

2.14 Nectar-gathering honey bee worker

23

place it at the pinnacle of insect evolution. There are several races of the honey bee throughout the world.

Honey Bee Biology

Honey bees are unique among insects in a temperate climate because they remain active throughout the year. During winter they cluster near the center of the hive and feed on honey to provide heat energy. When pollen and nectar become available, workers begin foraging and the queen resumes egg laying (queens may lay eggs through most of the year in tropical areas). As the season progresses and the colony reaches a peak of brood

2.15- Honey bee queen serves as the social center of the hive by producing chemicals called pheromones — and lays hundreds of eggs every day during the season.

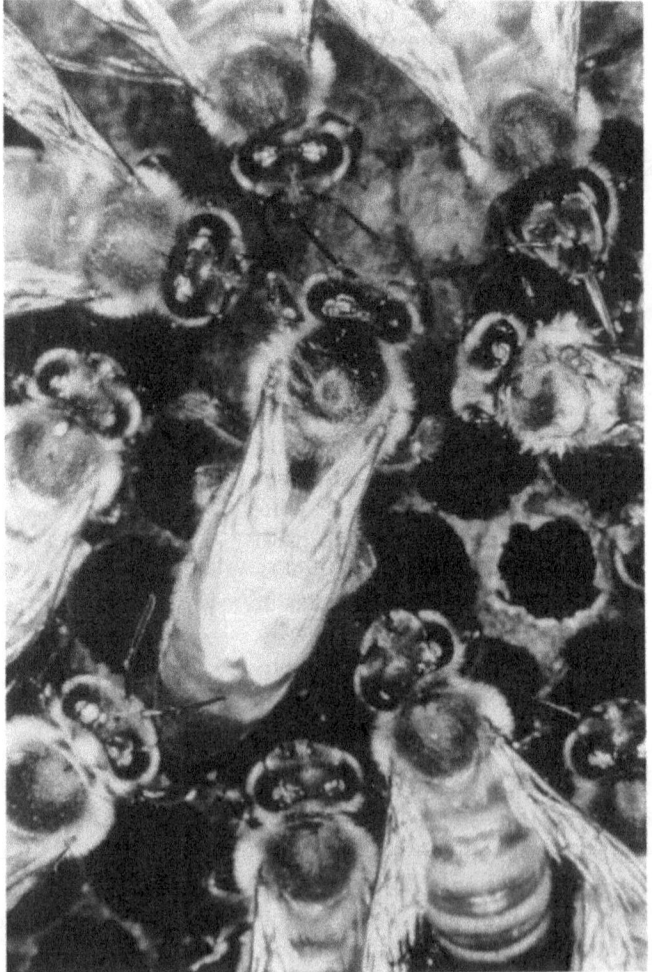

rearing activity, the queen may lay up to 1,500 eggs per day. For a variety of reasons, about 15% of the eggs never produce adults during summer. The peak production of new workers is about 1,275 per day.

During late spring, swarms will issue from large colonies, especially if there is insufficient hive space for brood rearing and honey storage, and if colonies are allowed to overheat on sunny days or crowd on rainy ones. The old queen and most of the younger worker bees fly out of the hive and form a swarm hanging from a nearby bush or tree. Previous to issuance of a swarm, workers will have prepared queen cells and started rearing new queens. The first young queens to emerge will either kill sister queens before they emerge, or fight to the death of all but one.

On warm, sunny days virgin queens will begin making 5 to 30 minute nuptial flights at 45-90 ft in the air. Typically, each queen mates with about 7 drones per flight and she will continue to make flights until she has at least 5 million sperm stored in her sperm storage organ. This usually requires mating with 12 or more drones. Queens may live for 1 to 3 years, but the number that run out of their supply of sperm to produce worker offspring increases with age.

Workers live 4 to 6 weeks during the active season

2.16-
Nurse
bee
feeds
larvae in
brood
comb of
the
honey
bee

which is divided into 2 to 3 weeks cleaning cells, feeding brood, building comb and guarding the colony and 2 to 3 weeks foraging in the field for pollen, nectar, water, and propolis (gum gathered from trees and processed to form brownish material to close openings in hive). Strong colonies contain 30-40,000 workers, a few hundred drones, and one queen at the height of the season. Pollen is a key feature of colony nutrition. It is eaten by newly-emerged workers for about 10 days and supplies the protein for royal jelly production. Royal jelly, in turn, supplies the precursor material for production of queen substance (pheromone) by the queen. Queen substance chemicals inhibit the construction of queen cells, attract drones during nuptial flights, and keep swarming workers near the queen. Reduction in queen substance in a colony leads to supersedure: whenever a queen is failing, workers attempt to produce new queens.

Alkali Bee Biology

The alkali bee is a soil-nesting solitary bee which occurs naturally in restricted arid areas west of the Rocky Mountains. In nature, nesting is confined to places where the soil is sub-irrigated over a hardpan layer which leads to relatively bare alkali spots.

Alkali bee adults emerge from the soil in late spring

2.17 - Alkali bee on alfalfa.

or early summer, depending on temperature and moisture of the soil. In general, males start emerging about a week before females. Mating activity is frequent at nesting sites where the males mate with females as they emerge from the ground.

Females begin constructing their nests soon after mating and prefer to dig in existing holes. The nest is a vertical shaft with a lateral tunnel that has oval cells branching from the underside. The first group of 6 to 8 cells is usually constructed at a depth of 4 to 6 inches. If conditions are favorable, the female will construct another series 8 to 12 inches deep in the soil. Average number of healthy larvae per female is 7 to 9 in managed nesting sites. Females live for 4 to 6 weeks, and under good conditions, are active for about 60 days.

Each cell is provisioned with an essentially pure ball of pollen (91%) moistened with a small amount of nectar. A single egg, that hatches in 2 to 3 days, is laid on top of each pollen ball. The larva consumes all of the pollen in 7 to 10 days and defecates to become the overwintering prepupa. The prepupal stage lasts 9 to 10 months until increasing soil temperatures of spring initiate the pupal stage.

Four basic conditions must be met and maintained in nesting sites to produce good numbers of alkali bees: (a) the soil is moist throughout the nesting area down to

2.18-
Alfalfa
Leaf-
cutting
bee
nest

a depth of 30 cm; (b) the soil is firm and compact without either a crusty or fluffy layer at the surface; (c) the surface is essentially bare with only sparse vegetation; and (d) the soil is a silt loam with 12 to 24% clay and 10 to 40% sand.

Alfalfa Leafcutting Bee Biology

The exotic alfalfa leafcutting bee is a solitary bee which nests in holes in wood or other materials. It winters as a mature larva (prepupa) in a cell formed from cut leaf pieces. Cells are constructed in series with females developing in the innermost cells of the tunnel and males in the outermost. Adults emerge from cells in late spring or early summer, depending on temperatures. A chilling period is needed to break diapause. Males emerge first, with a peak at about 3 days after first emergence. Female emergence peaks at 7 days. The male bee has mandibles with a prominent tooth adapted for chewing through the closing leaf plugs; while the female mandible has small teeth adapted for cutting leaf pieces. Females are not receptive to mating as soon as they emerge, but start building cells on the second or third day when sperm are present in their storage organs. Males cluster in nests or other cavities at night, but male populations dwindle quickly as females begin nesting. Females spend the night in the nest, faced inward. As temperatures rise in the morning, they turn around and face the entrance, but do not come out and fly until the temperature exceeds 68°F.

Females construct thimble-shaped cells of leaf pieces. After a cell is completed, the bee gathers nectar and pollen as food for the larva. The average provision mass is 64% nectar and 36% pollen. The female lays an egg on the surface of the nectar-pollen mass and caps the cell with round leaf pieces which also form the base of the next cell. She then constructs, provisions, and lays eggs in additional cells until the tunnel is nearly full. An entrance plug is usually formed from round leaf pieces.

BEE POISONING SYMPTOMS AND SIGNS

Honey Bees

Excessive numbers of dead worker bees piling up in front of the hive are the most common symptom and essentially a sure sign that chemical poisoning is involved. When poisoning is severe, thousands of bees will accumulate in front of a hive each day. Use of Todd dead bee traps on honey bee colonies has shown that up to 100 dead bees per day is a normal die-off, 200-400 is a low kill, 500-900 is a moderate kill and 1000 or more is a high kill. Also, 90% of the bees dying of old age die away from the hive. This symptom varies considerably with a number of factors. Strong colonies, slow-acting chemicals, and short distance to treated plants all lead to heaviest buildups. Speed of action of the chemical can make a sizable difference. For example, when plots were treated with fast-acting acephate (Orthene) as compared to slow-acting carbaryl (Sevin): even though Orthene affected colonies more severely, an average of only 3,387 dead bees were collected in front of the hives

Todd dead bee trap

during the first week after application. In the Sevin-treated plot, an average of 19,942 dead bees were collected in the first week. Dead honey bee workers which accumulate at the hive entrance usually represent 10-20% of the total number being killed. The rest of the poisoned foragers die in the field without getting back to the hive. However, with a fast-acting chemical, we have documented the proportion at the hive can be as low as 1%. When insecticide contaminated pollen is brought back to the hive, newly-emerged workers which feed on the pollen are also killed. Within a few days, the mass of dead and dying bees may be composed of greater num-

poisoned foragers

3.1 - A bee-keeper scoops up dead bees from an extensive pesticide kill.

bers of young workers dragged from inside the hive than of older workers getting back to the hive from the field.

Any type of poisoning is likely to cause the bees to become agitated and aggressive. When the hive cover is removed they fly off the top bars, sometimes straight at the beekeeper's head. This is especially evident with lindane and organophosphorus compounds. Aggressiveness is logically related to sociality of the honey bee which reacts to stress of the colony by stinging any

3.2 — Pesticide exposure often results in increased aggressive behavior by worker bees.

aggressive behavior

suspected marauder. Honey bees may become quite noisy following insecticide exposure, but they often produce loud angry sounds for a variety of reasons.

Stupefaction, paralysis, abnormal jerky, wobbly, or rapid movements, and spinning on the back are commonly caused by DDT, other organochlorine materials, and organophosphorus compounds. Bees are often observed performing abnormal communication dances on the horizontal landing board outside the hive while

stupefaction and paralysis

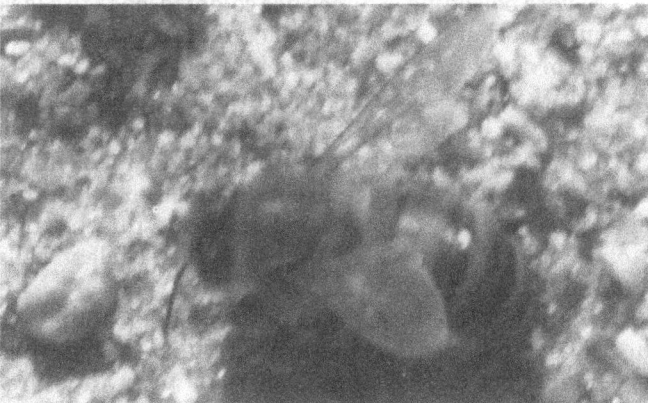

3.3 - A 'crawler' is unable to fly after insecticide exposure.

under the influence of chemical poisoning. Bees slightly affected by some organophosphorus compounds will crawl up the walls of the hive and fall to the floor over and over. Sublethal doses of parathion cause mistakes in communicating distance and direction of feeding sites and in time-sense. Disorganized behavior patterns may lead to lack of recognition of affected field bees by guard bees.

Many bees poisoned with Sevin (carbaryl) or dieldrin slow down, lose their ability to fly, and appear as

3.4 - A colony with poor house-keeping

though they had been chilled; such bees may take 2 to 3 days to die. These "crawlers" move about in front of the hive, but are unable to fly.

Severe poisoning of honey bee colonies leads to a lack of young workers: poor housecleaning is another sign typical of severe poisoning. The hive bees are unable to remove the dead bodies, and these may plug up the entrance so badly that the remaining foragers can hardly get through. Also, not surprisingly, all cleaning and general housekeeping are not tended to. Nectar is often deposited in empty brood cells and queens may

regurgitation

stop laying simply because there is a lack of clean cells to receive the eggs.

Regurgitation is a symptom which is unique to honey bees, because they live in a perennial colony. They are adapted to collecting quantities of nectar and preparing a sizable surplus of honey in order to survive the winter. The annual-nesting wild bees only collect nectar for their own survival and to feed their larvae. The provisions which the female alfalfa leafcutting bee places in each cell to nourish the developing larva contain about 64% nectar and 36% pollen; while the alkali bee provides a nearly pure pollen ball with 9% nectar for its larva. Daily consumption of honey syrup by honey bee workers and alkali bee and leafcutting bee females in laboratory feeding studies varies considera-

bly, with the honey bees ingesting 7 to 10 times as much per individual (50 µl compared to 5-7 µl). We thought the honey bee might logically have a larger proventriculus (honey stomach) than the wild bees. However, measurements showed this structure to be remarkably similar in the three species. External dimensions averaged 3.0 x 1.5 mm for the honey bee worker; 2.6 x 1.5 mm for the alkali bee female; and 2.6 x 1.4 mm for the alfalfa leafcutting bee female. Obviously, the difference in regurgitation is a behaviorial trait. Regurgitation of the proventricular contents by honey bees is especially associated with exposure to organophosphorus insecticides, but it also occurs with the synthetic pyrethroids. A wet, sticky mass of dead and dying bees accumulates in front of the hive. Also, these chemicals cause a large proportion of the workers to die with their glossae and par-

3.5 — Regurgitation of the honey bee's stomach contents often occurs after exposure to organophorphorous insecticides.

tongue extension

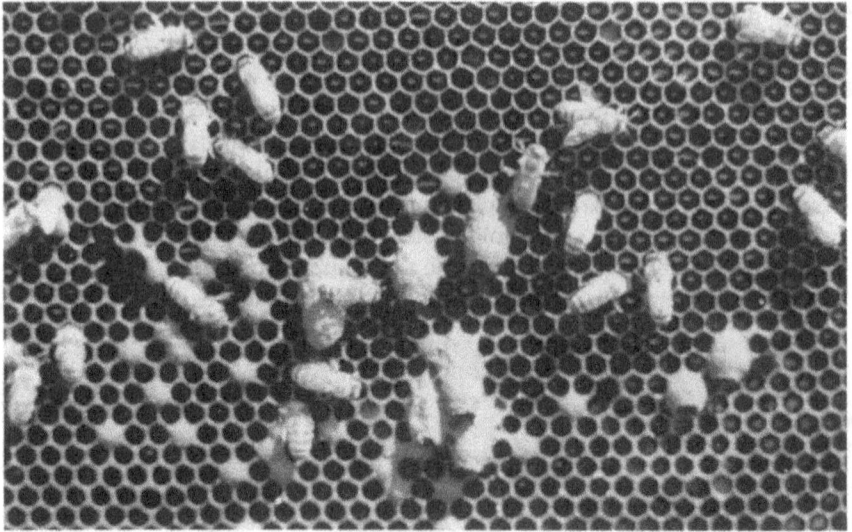

3.6 - Emergency queen cells appear after a queen is killed.

aglossae (tongue) extended: 99% with organophosphorus poisoning; 69% with synthetic pyrethroids; and 42% with other chemicals and in normal die-off. About 45% of leafcutting bees and 40% of alkali bees die with their tongues extended regardless of the cause of death, and they remain dry, not soaked with regurgitation. With the exception of materials causing excessive tongue extension and regurgitation by the honey bee, there is no significant difference between the three species.

3.7 — Queenless colonies often produce drone brood.

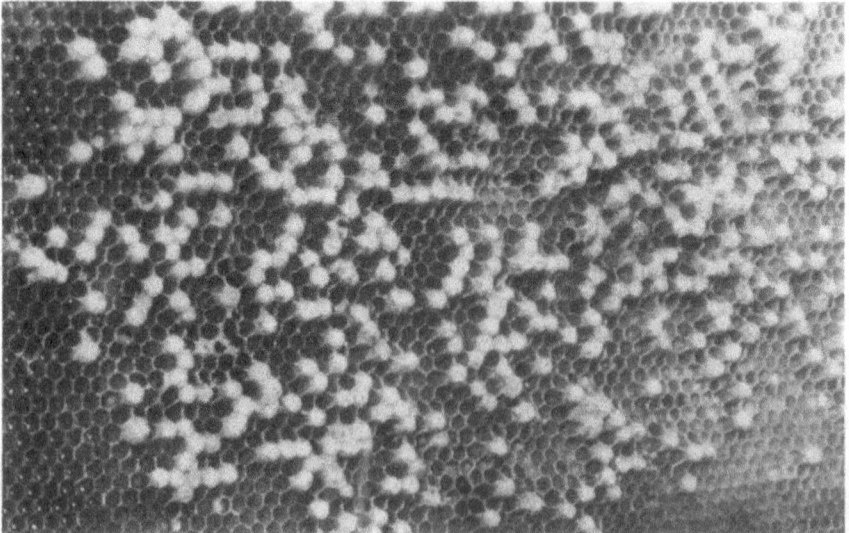

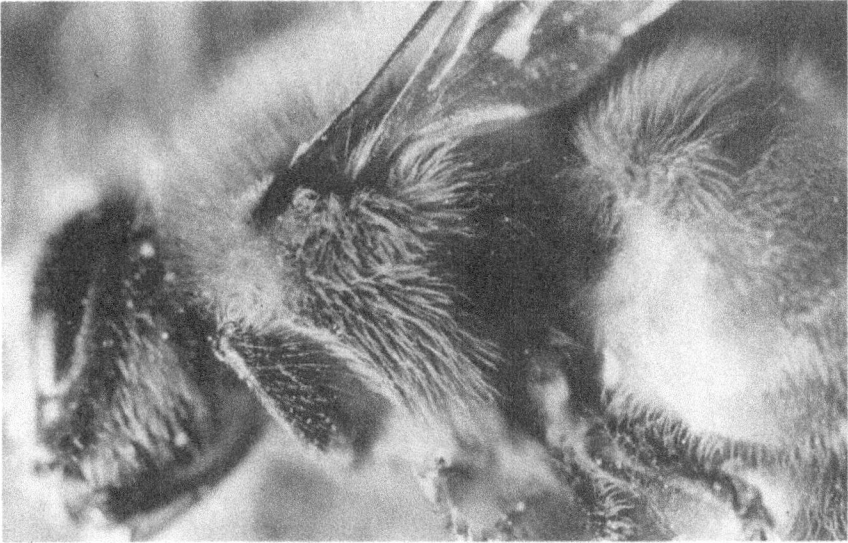

A basic difference between the social bee and the solitary bee is the exposure of reproductive stages to pesticides. The egg-laying female honey bee is never directly exposed to insecticide residues; while the egg-laying females of alkali and leafcutting bees have the greatest exposure because they are the active foragers. Indeed, apparently healthy honey bee queens are often seen amidst the handful of remaining workers several weeks after the onset of a severe chemical kill. Usually,

3.8 — Insecticide particles and pollen on bee's body.

3.9 — Young bees feed on this pollen and die

the queen will be superseded within the first 30 days following such a kill. This is a clear indication of contaminated pollen, elimination of newly emerged workers, and reduction in queen substance.

Loss of hairs on honey bee workers is associated with arsenical poisoning.

Pesticides most commonly contaminate the interior of the hive when they are carried to it in the pollen loads of the foraging bees. Workers can also carry lethal concentrations of insecticide back to the hive in their honey stomachs. If dead and dying light-colored, newly emerged workers are seen, it is a sure sign of pollen contamination. Newly emerged adult bees feed actively on pollen. Such pollen can remain toxic to bees after storage in the combs for up to 8 months (Sevin) or even a year (Penncap-M). When honey bee colonies have been severely poisoned, dead brood may be found in the combs or pulled out or crawled out onto the bottom board or in front of the hive. Individual larvae are most often killed by desiccation and related starvation. Chilling is not the usual cause of death during the insecticide application season. When not enough hive bees are left to cover the brood frames or care for the brood, desiccation or starvation kills the larvae and pupae.

Queens may be affected by contaminated pollen or nectar and behave abnormally; for example they might lay eggs in a poor pattern or even be killed. Though again at times, apparently healthy queens are seen amidst the handful of remaining workers several weeks after the onset of a severe chemical kill.

supersedure of
the queen

Queens are often superseded or the colony becomes queenless. This is what we suspect is involved when the brood cycle is broken within a few days of the time the insecticide was applied. Foragers have probably stopped bringing in pollen. When there is a lack of pollen, the hive bees will begin feeding on the eggs or there may simply be a lack of clean and/or empty cells for the queen to lay in because foragers are depositing nectar in the brood combs. The queen often remains alive and apparently healthy for a week or more before she is superseded. Under normal conditions, healthy workers begin to produce royal jelly (the queen substance precursor) a few days after they emerge. Lack of brood is not related to supersedure, but the supply of queen substance is, and when it falls below a critical amount the workers will get rid of the queen: the contaminated pollen-queen supersedure syndrome. When there are no eggs or young larvae present, the workers can no longer rear a new queen and no supersedure cells are present. Queenlessness has been associated with the use of a wide variety of insecticides, including arsenicals, dieldrin, Sevin, Orthene, malathion, parathion, and Penncap-M. Typically, severe Sevin dust poisoning makes at least half of the colonies queenless within 30 days. Severely weakened and queenless colonies do not survive the following winter.

queen sterilization

Dimilin (diflubenzuron) can cause the development of a drone layer if it is fed to the colony at a relatively high dosage rate. We suspect the queen becomes functionally sterilized because of the effect of the chemical on her sperm storage and transfer structures or egg covering structures. Her eggs are no longer fertilized as they pass through the oviduct.

Honey bee workers may stop bringing pollen back to the colony within the first day of exposure to severe insecticide poisoning. Studies in the forest near LaGrande, Oregon, showed this clearly when we only obtained pollen trap samples the first day after Orthene and Sevin treatments which severely damaged and ultimately killed the test colonies. Dimilin applications did not affect the bees and they continued to collect pollen for the duration of the investigation.

Lack of bees foraging on a crop that normally is attractive to bees is another sign of bee poisoning. Bees visiting the crop are killed by the pesticide. If large numbers of bees are killed, many dead or dying bees may be found on the ground. Depending on the chemical involved and the crop, it may take up to 7 days or more before bees begin to visit the treated field again.

Alfalfa Leafcutting Bees & Alkali Bees

A distinctive symptom of poisoning in leafcutting bees and alkali bees is simply a lack of nesting females in the field shelters and at the soil nesting sites. Large masses of dead and dying bees in front of the hives, which are the most common sign of poisoning in honey bees, are seldom seen with either wild bee. This differ-

survival of male
alkali bees

ence is easily understood when you compare the numbers of foraging females per nest area. A realistic comparison of healthy populations of the 3 species provides a ratio of one alkali bee/32 leafcutting bees/572 honey bees/ft² of nest area.

An alkali bee bed without females will often have a number of males milling around above the surface in typical circling flights for several days after the poisoning occurred. Males form sleeping aggregations (roost) at night on weed stalks in field edges and waste areas near the beds. Their daytime activity is mainly involved with patrolling flights in nearby fields or around the bed searching for females. Since the females commonly forage up to a mile or more from the nesting site, they can be killed by insecticidal residues which are never contacted by the males. Night insecticide applications would seem to be a special hazard to males roosting nearby, but we have not been able to document that significant losses occur in this way.

A seemingly dramatic decrease in female leafcutting bees which is assumed to be caused by insecticide poisoning sometimes proves to be the natural lull between

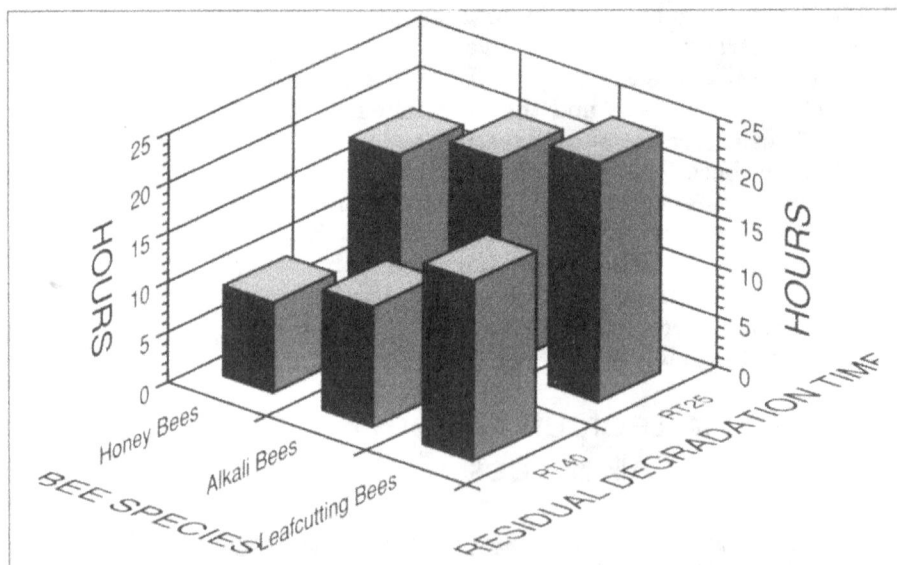

3.10 — Larger honey bees respond faster after chlorpyrifos treatments than do alkali bees or leafcutting bees. The RT_{40} and RT_{25} values indiicate the time required to bring bee mortality down to 40% and 25% in cage test exposures to field-weathered spray deposits.

cycles. When the bee larvae are incubated at 85-90°F, the adults emerge during a short period of time. What appears to be good bee activity 5 weeks after the bees are placed in the field may be reduced to near zero at 6-7 weeks.

Some persons have associated a lack of carrying either pollen or leaf pieces by leafcutting bee females or pollen by alkali bee females with insecticide poisoning. However, we believe this behavior is caused by old age or senility, since it is observed beginning about 4 weeks after the start of female nesting activity in the field regardless of any insecticide exposure.

About 45% of leafcutting bees and 40% of alkali bees die with their tongues extended regardless of the cause of death, and they remain dry, not soaked with regurgitation.

Aggressiveness and stinging is not typically associated with pesticide poisoning of wild bees. The solitary leafcutting and alkali bees rarely sting under any circumstances.

General symptoms of insecticide poisoning, such as wobbly and uncontrolled movements, are similar for honey bees, alkali bees, and alfalfa leafcutting bees. However, wild bees exposed to Sevin (carbaryl) become hyperactive. During the final stages of intoxication with any insecticide, leafcutting bees often spin on their backs much like house flies treated with DDT and alkali bees perform numerous rapid turning motions while lying on their sides. All three species become highly agitated and fly wildly following exposure to Temik (aldicarb).

Another symptom which is detectable in controlled cage studies is the production of strident sounds by the alkali bee following insecticide exposure. This species is normally quiet except for the low buzzing sound emitted during flight.

Onset of poisoning symptoms is delayed when the 3 species are given low dosages of chemicals in chronic feeding studies. Lowest dosages which cause greater mortality than the untreated checks typically require a week or more for development of detectable symptoms. Any expression of symptoms in long-term feeding studies is associated with some level of mortality greater than the checks.

Mortality curves from such tests which attain less

than 100% mortality will reach a plateau and continue essentially unchanged through the 21st day. This is the result of killing only the more susceptible individuals in the test group.

A final difference in reaction of the three species is the typical relationship of susceptibility to insecticides in the field. Since essentially all cases of poisoning occur from chance adherence of insecticidal residues to bees foraging treated plants, the main reason for greater susceptibility in the leafcutting bee is probably its greater surface/volume ratio. Surface/volume ratio for the leafcutting bee is twice that of the honey bee and for the alkali bee, 1.3 times that of the honey bee. Clearly, greater susceptibility of the alfalfa leafcutting bee is related to the larger relative amount of toxicant it accumulates on its body surface.

Summary

Specific differences in reaction to insecticide treatments between the honey bee, the alfalfa leafcutting bee and the alkali bee are sometimes related to differences in their sociality. Increased aggressiveness, regurgitation of honey stomach contents, and accumulation of dead and dying workers at the hive are unique to the honey bee because it lives in sizable social colonies. Why Sevin (carbaryl) causes a marked reduction in honey bee activity as compared to intense hyperactivity in the wild bee species is not as easy to explain. Greater susceptibility of the diminutive leafcutting bee to most insecticides appears to be related to its surface/volume ratio which is twice that of the honey bee worker.

TYPES OF PESTICIDES   4

Nature of Pesticides

The term pesticide was coined about 35 years ago as a collective noun to cover all materials used to control, destroy, or mitigate pests. In common usage, a pesticide is any chemical used to harm living things harmful to man — pests. But, bactericides are not normally lumped with other pesticides.

Before pesticides, insects and diseases destroyed part of every crop every year. Farmers adjusted and sowed a portion of their crop for pests, planted part for the pests and part for themselves.

In the modern world and competitive modern agriculture, pesticides are important tools. They've helped make food cheap and easier to grow and store. They've saved millions of lives by destroying harmful insects such as mosquitos. Humans live longer thanks to modern pesticides. Pesticides are part and parcel of modern food production and health standards. Pesticides kill by:
- contact
- stomach poison

Pesticides kill pests. That's their job. To destroy and control living things.

• fumigation

Some kill by only one action while others kill by various combinations of 2 or all 3 actions.

Use of pesticides creates some problems and they do have limitations. Problems occur when pesticides are applied:

• in the wrong manner
• at the wrong time
• at the wrong place

killing honey bees or wild bees. A problem called bee poisoning. Bee poisoning is the accidental killing of bees by pesticide applications. Most bee kills are not intentional, but result from accidents or mistakes. Also, there may be situations when it is necessary to deliberately kill bees with a pesticide. An example is northern beekeepers who kill their bees in the fall rather than overwinter the colonies which is not bee poisoning. But *most* bee poisoning incidents occur from applications of agricultural pesticides.

Kinds of pesticides are:

• rodenticides
• fungicides
• miticides
• herbicides
• insecticides

Only insecticides cause serious bee poisoning. Rodenticides never cause bee poisoning.

Fungicides control fungi. Fungus parasites attack many agricultural crops. They are lower plants lacking green chlorophyll and unlike plants, they cannot make their own food, but are parasites of green plants or animals. In general, fungicides are not harmful to bees though several of the mercury-containing compounds appear toxic to bees. Also, captan may be somewhat toxic to honey and leafcutting bee larvae.

Herbicides control weeds; a plant growing in the wrong place at the wrong time. Much the same as what happens to a beekeeper involved in an accident — showed up at the wrong place and time. Most herbicides do not harm bees, though there are exceptions such as the arsenical and dinitro compounds.

Miticides control mites; small animals related to insects. They differ by having two body parts, eight legs and never have wings. Also, most mites are small beasties barely visible without a hand lens. Most specific miticides do not harm bees. Pesticides that kill both mites and insects may harm bees.

Insecticides control insects; animals with 3 body parts, 6 legs, and many have wings. Only a small part of the million different species of insects that abound on earth are pests. These pests harm us by eating our crops, gardens, roses, shade trees, and transmitting plant, domestic animal, and human diseases. Since a bee is an insect, it comes as no surprise that insecticides are harmful to bees and also the pesticide of most concern to beekeepers.

There are 4 major groups of synthetic insecticides:
- chlorinated hydrocarbons (organochlorines)
- organophosphates (organophosphorus)
- carbamates
- pyrethroids

The toxicity of the insecticides from and within the groups to bees ranges from non-toxic to highly hazardous (see Appendixes II through VI). There will be more about the different groups later. Ideally, the insecticide selected for controlling a particular pest should not harm bees. It's necessary to find the toxicity to bees of the particular proposed pesticide. In Appendixes II through V pesticides have been divided into 4 toxicity groups according to their toxicity to different bees (at their usual application rates) and the length of time they remain toxic after application.

In Appendix VI, technical information and explanations of the LD_{50}, RT_{25}, and RT_{40}, and application rates are given to fine tune an insecticide program without harming bees.

5   HERBICIDES

Herbicides kill plants — a completely separate kingdom from insects and bees. Unlike insecticides, most herbicides do not affect the nervous system of animals. Except for some general all-purpose killers, most herbicides present only a low hazard to bees. The elimination of valuable nectar and pollen plants is the major effect of herbicides on bees. For this reason, and the massive impact on bees from the loss of food plants, broadspectrum herbicides are not justified on wastelands, roadsides, or railway embankments.

Herbicides which affect bees

Herbicides have generally been assumed to be low in hazard to bees, with a few exceptions such as the arsenicals, dinoseb, and Endothal. Aminotriazole, Atrazine, and Simazine are low in toxicity but present some hazard because treated flowers remain open, allowing residual action to occur.

In sugar syrup feeding studies (when toxic compounds are added to sugar syrup and fed to bees in cages) the most toxic materials tested on bees were paraquat and the organic arsenicals MAA (methylarsenic acid), MSMA (monsodium methylarsenate), DSMA (disodium methylarsenate), hexaflurate and cacodylic acid (organic arsenicals). They were also toxic when sprayed on caged bees. Phenoxy materials (2,4-D; 2,4,5-T; 2,4-DB;

silvex) were relatively nontoxic, except at unusually high dosages in feeding studies. At those rates, brood rearing in colonies was severely reduced as long as the materials were being fed to the bees.

However, phenoxy herbicides may exert adverse effects on bees in the field that are not detected in most bee poisoning experiments. Such effects might be caused by poisoning of the nectar and by reduction of the bee's ability to fly. It has been suggested, that since phenoxy herbicides alter plant secretions and metabolism, a change in nectar composition of plants sprayed with hormone weed killers would be expected. This change could harm bees. Heavy bee losses have been reported with Tormona 80 (a 2,4,5-T salt) when used at 6 times its normal concentration in wooded areas close to bee forage.

The phenoxyacetic herbicides such as 2,4-D may inhibit nectar secretion even when used at concentrations that cause no visible plant injury, and less nectar is then available to bees.

Some mixtures of two or more insecticides do not cause greater kills of bees than the same materials applied singly. However, a mixture of dimethoate (Cygon) insecticide and 2,4-DB herbicide was less hazardous than Cygon alone. Mortalities from diazinon, malathion, methyl parathion, parathion, and mev-

5.1— Herbicides reduce bee forage, and a few may harm bees, however, elimination of blooming flowers from orchards and fields often protects bees.

inphos were substantially reduced in some combinations with herbicides. In another test, insecticide-herbicide mixtures fed to honey bees were not more toxic than the insecticides alone. Use of blossom thinners, dinitrocresol, Amid-Thin (naphthalene acetamide), NAA (napthaleneacetic acid), Alar, and ethephon in orchards have not been hazardous to honey bees. Nor has use of xylene, diquat, and Acrolein for control of aquatic weeds in irrigation ditches and canals harmed bees collecting contaminated water.

Even excessive dosages (up to 1200 lb acre) of soil sterilant herbicides, usually containing mixtures of sodium borates and sodium chlorate, have not been hazardous to either adults or larvae of the alkali bee in soil nesting sites.

Desiccants applied to alfalfa seed crops or cotton, usually formulated with dinitrocresol, endothal, or sodium chlorate, have not caused bee poisoning problems.

Summary

The damage to bees from herbicides is much greater as a result of destruction of the bees' food sources than as a result of direct poisoning.

INSECTICIDES 6

Organisms of any kind may be killed mechanically, physically, or chemically. Insecticides kill chemically by reacting with a body component causing death by poisoning. Insecticides fall into six major groups. Natural products used for insect control are (1) inorganic materials and (2) botanicals. Synthetic organic compounds include (3) chlorinated hydrocarbons, (4) organophosphorus materials, (5) carbamates, and (6) pyrethroids. These compounds are synthesized by humans in the chemistry lab. In addition, there are several miscellaneous groups of insecticides. Although several synthetic organic insecticides were known previous to World War II, not until the discovery of DDT did interest in these chemicals become widespread. Nearly all synthetic organic insecticides now in use were developed after 1947.

"Miracle" Insecticides: The Chlorinated Hydrocarbons

The organochlorines represented a major breakthrough in our war with insect pests, allowing us, for the first time in history, to nearly eliminate crop pests. DDT is the best known of the group. First made in 1874, its insecticidal effectiveness was not discovered until 1939 by Geigy: it was patented in 1942 at the same time the insecticide lindane was discovered. DDT is one of the most important insecticides ever devised by man. It has saved millions of lives through the control of insect vectors of human diseases such as malaria. Also countless humans were saved from starvation. Chlorinated hydrocarbons include DDT and its analogs, cyclodienes and hexachlorocyclohexanes.

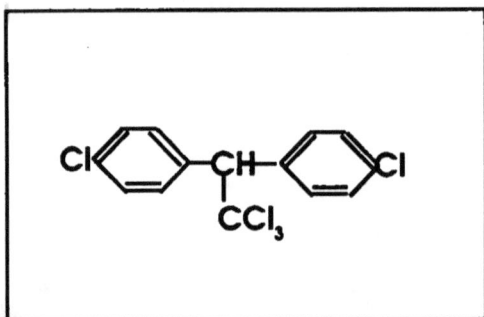

6.1 - DDT molecule.

All organochlorines are molecules composed of chlorine, hydrogen, and carbon, and occasionally oxygen or sulfur. We know painfully little about how these poisons actually kill insects, except to say they attack the nervous system. The group, as a whole, is persistent when applied to plants or other surfaces. In soils they give long residual action against insect pests. For example, chlordane kills termites in the soil 20 years after application. Chlorinated hydrocarbons are not systemic or translocated in plants. They are broad spectrum insecticides, killing nearly all insects contacted. However, five of the diphenyl DDT relatives such as Kelthane and chloropropylate, are specific miticides. There are real differences between the compounds and their effects on bees.

The cyclodiene group of organochlorines (chlordane, dieldrin, aldrin, heptachlor) and lindane tend to have a residual toxicity highly hazardous to bees. Exceptions with short residue actions are endrin, Thiodan (endosulfan) and toxaphene.

On the other hand, DDT and similar compounds

(TDE, methoxychlor, Perthane [ethylan]) tend to be moderate in bee toxicity when applied as sprays. They can be used safely while bees are not foraging. Honey bees are insensitive to external DDT, but are very sensitive to injected DDT. This difference is commonly attributed to barrier effects of the skin or cuticle. Also, in this toxicity pattern, bees are unlike other insects and more like mammals which are also insensitive to external DDT. The organochlorines have traditional chemical names since they were developed before modern marketing methods. Unlike more modern chemicals they are not going to Ambush, Pounce or Prowl on us. The zenith of the chlorinated hydrocarbons occurred from the late 1940's to the early 1960's. Use, development, and manufacture of chlorinated insecticides dwindled with the years. While justified for other environmental criteria the disuse of this group has not always been good for bees. The insecticides often fitted into pest control programs presenting little bee hazard.

*Symptoms: Typical bee poisoning symptoms from **chlorinated hydrocarbons**: trembling, erratic activity, dragging hind legs, wings hooked together and held away from body, many bees able to fly, many bees die in field as well as at the colony.*

Swords Beaten into Plowshares: The Organophosphates

During World War II, both sides devoted efforts to preparing organophosphates (nerve gases) for warfare. The Germans discovered the insecticidal action of these compounds during the war. Gerhard Schrader in Germany discovered the toxic effects of these compounds on insects while looking for a substitute for the insecticide nicotine. Organophosphates (OP) represent a case where a sword was turned into a plowshare. The first OP, Bladan, contained TEPP as the active ingredient. In 1944, parathion was introduced. It and its relative, methyl parathion, are the most widely used organophosphates today.

cholinesterase, a vital enzyme, inhibited by organophosphates

The term organophosphate is a generic term used to cover all the toxic organic compounds containing phosphorus. They kill animals, including insects, by inhibiting cholinesterase, a vital enzyme of the nervous system. Constant disruption of nervous activity occurs at the nerve endings. Insects literally jump their nerves to death.

Organophosphates are characterized by their effects on insects. Some, like Lorsban (chlorpyrifos), have long residual action, while others such as TEPP have short residual action. Some kill only certain insect species, while others kill all insects contacted. Some, like Systox (demeton), are systemic. Applied to the soil or plant, the toxicant translocates through the plant to kill pest insects feeding on the plant. Some, like TEPP, have high volatility or fumigant activity, and may be absorbed through the bee's spiracles or respiratory system. TEPP reacts with water quickly and is very nonpersistent and dissipates rapidly. This diversity carries over to bee poisoning. The OP's fall into 3 broad groups:

- highly toxic to bees
- highly toxic with short residue activity
- relatively non-toxic [Dylox (trichlorofon), schradan]

As a general rule, many OP's are highly hazardous to bees and cannot be applied safely to flowering crops. Examples are Baytex (fenthion), diazinon, parathion, and Guthion (azinphosmethyl).

Because of short residual action, some organophosphorus materials such as TEPP, Trithion (carbophenothion), Dibrom (naled), phostex, delnav, korlan, and menazon can be applied safely to flowering crops during the evening or early morning when bees are not foraging. Plants take up systemic OP's such as Thimet (phorate), Systox (demeton), Di-Syston (disulfoton) and schradan from the soil or through their leaves. Generally, these present little danger to bees. Bees don't contact the poison if it is applied when they aren't foraging. Also, Systox repels honey bees. Some of the systemics such as Metasystox R (oxydemetonmethyl) retain a longer hazard to bees when applied to old, senescent foliage. Some OP's, such as isopropyl parathion, are 250 times more toxic to flies than to bees. This demonstrates the target selectivity of this group of insecticides.

The greater brain acetylcholinesterase (AChE) concentration in young honey bees gives them greater tolerance to malathion, as compared to older bees. Reductions in AChE correlate with development of organophosphate insecticide poisoning symptoms. Also, in bees, different species have different AChE specificity. The alfalfa leafcutting bee has an unusually high tolerance to Dylox (trichlorfon) when compared to other species of bees. Only the relatively high pH of the body fluids as compared to that of honey bees correlated with this specificity.

The organophosphates are often the most toxic to honey bees with parathion (LD_{50} = 0.18 µg/bee) high on the list.

Organophosphate insecticides out number other groups. Over 50,000 different OP pesticide compounds exist. They account for about 25% of the registered synthetic insecticides in the United States. For many years they were the most popular group of insecticides.

The popularity of the organophosphates in agriculture is declining. Insect pests have become resistant to some compounds. Many organophosphates are highly toxic to humans and other non-target organisms, patents have expired for others, and some are just no longer profitable for companies to produce. So production of many OP's has stopped.

*Symptoms: Typical bee poisoning symptoms of **organo-phosphates:** regurgitation, disoriented, distended abdomens, erratic, wings held away from body and hooked together, tongues extended, many bees die at colony.*

Trial by Ordeal:The Carbamates

In West Africa during the 1800's suspects in criminal cases were forced to consume beans of a poisonous plant, *Physostioma venenosum.* If the suspect survived, he was assumed to be innocent—if not, verdict and punishment were accomplished simultaneously. The nature of the poison aroused interest in Europe and in 1864 eserine, the active principle, was isolated. Later, medicinal carbamates were developed from eserine and its analogs. In 1947, Geigy of Switzerland developed insecticidal carbamates. In 1957, Sevin (carbaryl), the most popular carbamate, was described by Geigy. Carbamate insecticides are generally aromatic compounds and methyl or dimethyl derivatives of carbamic acid. For the most part, these materials biodegrade easily and don't constitute the residual hazard of the chlorinated hydrocarbons. Carbamates kill insects and mammals entirely by cholinesterase inhibition at the nerve endings. Their killing action resembles that of the organophosphates.

Carbamates typically show erratic patterns of selec-

Symptoms: Typical bee poisoning symptoms of carbamates: erratic, stupefaction, paralysis, break in brood cycle, queens cease egg laying, supersedure queen cells, most bees die at colony; Sevin specifically causes inability to fly (crawlers).

tive toxicity to insects. In general they are not broad-spectrum insecticides and their residual action is short. Some, like Temik (aldicarb), are systemic insecticides. As with other insecticides, formulations vary from granule to liquid. Bee hazard of the insecticides in this group varies also. Honey bees are very susceptible to six carbamates. The probably source of this extreme susceptibility to these phenyl carbamates is the very low level of phenolase enzymes in honey bees. In other insects, these enzymes reduce the effects of carbamates by detoxifying them. Also, carbamates show consistently more toxicity to honey bees than to house flies. Sevin (carbaryl) is a striking example of an insecticide essentially non-toxic to warm-blooded animals but toxic to bees. Worldwide, Sevin is the most popular and widely used carbamate. Early laboratory tests indicated that Sevin was low in bee toxicity, but later field tests showed it to be very dangerous to bees. Residues may continue to kill up to 12 days or more after application.

The systemic carbamate, Temik (aldicarb) displays the reverse of Sevin (carbaryl) being highly toxic to warm-blooded animals, mites, and insects. Properly used as a systemic, it is non-hazardous to bees. However, Temik, injected into the soil several weeks prior to bloom in carrot seed fields, repels honey bees from the fields.

Some carbamates are the most toxic to honey bees with Furadan (carbofuran) (LD$_{50}$ = 0.6 µg/bee) high on the list.

New is better and this philosophy sums up much of the development of applied insecticidal work. Always there's been a new insecticide to replace discarded old ones. This is true even for large groups of insecticides. First the chlorinated hydrocarbons followed by the organophosphates and the carbamates: many lump these as first generation insecticides. Now, the bright stars are the second and third generation products; pyrethroids (synthetic pyrethrums) and insect growth regulators.

The Second Generation: Pyrethroids (Synthetic Pyrethrums)

Pyrethroids are gaining favor for insect control.

More new compounds are developed each year. These pyrethroids are related to the natural pyrethrum, but are synthesized from petroleum. The first practical compound of the group, allethrin, was synthesized in 1949. But, few pyrethroids came into use for pest control until the late 1970's. Resmethrin, first discovered in England, has low toxicity to mammals and is toxic to insects. It's used mainly in as a household insecticide in aerosol spray cans to control flying insects. Direct toxicity to bees is high though residue activity lasts only a few hours. Resemthrin is registered for killing honey bee colonies. It has an affinity for beeswax and honey and should be used only when combs are to be destroyed.

Symptoms: Typical bee poisoning symptoms from **pyrethroids:** *regurgitation, erratic, paralysis, many bees die between foraging area and colony.*

Synthetic pyrethrums now widely used in agriculture are fenvalerate (Pydrin, Ectrin) and permethrin (Ambush, Pounce, Ectiban). These materials have high insecticidal activity and are resistant to breakdown by sunlight. Many others are being developed. Some examples are cypermethrin (Cymbush, Ammo), flucythrinate (Pay-Off), and fluvalinate (Mavrik, Spur, Apistan). These extremely potent insecticides kill insects at very low dosages. For example, one recommended rate per acre for Spur (fluvalinate) is 0.12 lb AI/ acre. Actually, this is like spreading a quarter cup of liquid over an entire acre of land.

Some pyrethroids are broad spectrum insecticides while others kill only certain groups of insects. All except fluvalinate kill bees when sprayed directly on them. In general, their residual hazard is high to bees, though there are exceptions and variations. For example, Pydrin (fenvalerate) is moderately hazardous while fluvalinate (Mavrik, Spur, Apistan) and trulomethrin (Scout) are essentially non-hazardous. Fluvalinate also shows potential as a control for varroa mites attacking honey bees. Some, like Pounce (permethrin), when applied to flowering plants reduce the number of bees visiting that crop to near zero. Some like Pydrin cause only a partial drop in number of foragers. We don't know how the materials cause the reduction of bee foragers. Two possible ways:

• Sub-lethal poisoning changing the behavior of scout bees reducing recruitment of other bees
• A change in flower odor

The Third Generation:
Insect Growth Regulators

Insect growth regulators (IGR's) interfere or alter the normal growth patterns of insects in various ways and indirectly cause the insect's death. Several IGR's are synthetic analogues of juvenile hormone, a material governing the metamorphosis of insect larvae. Others are not, but still appear to interfere with natural insect hormones. IGR's don't kill adult insects, but kill the young as they develop or when they shed their exoskeleton and molt. Only a few IGR's are presently available for insect control, but they appear to be effective on a wide range of insects.

Honey bee larvae fed juvenile hormone, or treated topically with it, were recognized as abnormal by the nurse bees, and removed from the hive; brood mortality can be as high as 100%. If treated larvae survived to become adults, they had completely degenerated food glands, and were unable to rear brood.

Some IGR's like Enstar (kinoprene) which interferes with reproduction of Hemiptera (true bugs) kill only certain groups of insects. Some, like Insegar (fenoxycarb) have a broad spectrum of activity against different insects. Both appear to be harmless to bees.

Altosid (methoprene) received full registration in the United States in 1975 and thus became the first successful IGR for commercial use. It is used to control floodwater mosquitoes and does not harm bees. Dimilin (diflubenzuron) is a urea compound registered for control of insect pests on several crops. It inhibits cuticle (skeleton) formation, killing larval stages of many kinds of insects. It is favorably selective and is low hazard to bees and other beneficial insects. Dimilin fed to honey bee colonies in high concentrations in sugar syrup can cause queens to become drone-layers, supersedure, death of larvae and production of undersized workers. However, in 1976-77 studies, no adverse effects of Dimilin were noted on either brood or adult bees when bees collected pollen and nectar from plants treated in large-scale field trials. It had no adverse effects on either brood or adult bees. In more recent years, Dimilin has been widely tested for bee hazard and is safe to bees at normal dosages.

Inorganic Insecticides

Many inorganic materials have been employed in insect control, including compounds of antimony, arsenic, boron, fluorine, lime-sulfur, mercury, selenium, sodium fluoride, sulphur, and thallium. Some other early inorganic insecticides were common materials such as ashes, soot, and road dust.

Only the arsenicals, particularly lead arsenate, are still commonly used insecticides. Paris green, an arsenical insecticide, was first used in the United States in 1867.

Arsenicals are very toxic with a long residual action against honey bees (See Chapter 1).

Oils

Oils in their natural state are highly phytotoxic to plants, but when refined and used in an emulsion they may, under certain conditions, be applied safely to plants to kill insect pests. They are usually used in the winter or early spring when plants are dormant, although there are summer oils. These same conditions also reduce possibilities of bee kills. There have not been any reported bee losses from the use of oil insecticides.

Dinitro Compounds

Dinitro insecticide compounds were introduced in 1892 and reached peak popularity during the 1930's and 1940's. Only a few still exist as pesticides.

DNOC is 2.5 times more toxic to honey bees than to silkworms. Dinoseb (dinitro) is a general pesticide highly hazardous to bees. Dinitro compounds with low bee hazard are Karathane (dinocap) and Morocide (binapacryl)

Botanicals

Organic insecticides from plant products have several uses in insect control. These botanicals come from living plants and along with inorganics, are man's oldest pesticides. Botanicals are no longer widely used in agriculture. Compared to synthetic organics, they are higher priced and lack the long residue activity of the synthetics. Nicotine, pyrethrum, and rotenone are the most important botanicals.

The direct toxicity of the group to bees is high, but all have less than a few hours residual action.

Nicotine

Nicotine was first used as a concoction in 1690 against the pear lace bug in France. In America, beginning in 1746, infusions of tobacco leaves were used as an insecticide for plum curculio. Now, nicotine is obtained from tobacco commercially by steam distillation or solvent extraction. Nicotine kills insects rapidly, often within hours. Evidence indicates nicotine acts on the nerves of insects. Black Leaf 40, containing nicotine, is still a common insecticide.

Nicotine is much more toxic to honey bees than to mealworms, Japanese beetles, and cockroaches. Direct toxicity of nicotine to bees is high though the residue action is only a few hours. At present, nicotine is not used in situations where it may harm bees.

Rotenone

Rotenone is found in 68 species of leguminous plants. Since 1848, a variety of native roots from plants like *Derris* and *Tephrosia* which contain the poison rotenone have been used as insecticides. These roots are also used as fish poisons. Sometimes these insecticides are called derris dust or cube root. Rotenone has low toxicity for warm blooded animals. It kills by inhibiting respiratory metabolism and blocking nerve conduction. It is a selective insecticide; very effective against some insects and almost harmless to others. For example, rotenone is 1000 times more toxic to honey bees than to cockroaches.

It is toxic to honey bees, but residues last only a few hours.

Ryanodine

Ryanodine is the active insecticide ingredient from plants belonging to the genus *Ryania*. Ryanodine is toxic to fish, but not to warm blooded animals. The ground-up stems of Ryania are used as an insecticidal preparation, commonly under the name of Tyanex.

It is toxic to bees, but residues last only a few hours.

Pyrethrum

The insecticidal activity of pyrethrum was discovered in Iran around 1800. Great secrecy surrounded the source of the material, which was called Persian Powder, and it sold at extravagant prices. Pyrethrum was introduced into the United States about 1858. It has been widely used and is still the most popular botanical.

Pyrethrum comes from daisies (*Chrysanthemum* spp.). It is produced by grinding the flowers and mixing them with a dust diluent. Or, more commonly, by extracting the active toxicant from the flowers with solvents and formulating the extracts into sprays.

It is still widely used, particularly around homes because of its remarkably low toxicity for mammals. It is a spectacular insecticide; knocking down insects almost instaneously, but it has the distressing property of permitting total recovery of the victim in some circumstances. Pyrethrum breaks down rapidly, leaving no harmful residue. It is no longer used much in agriculture, but still used widely as a household and backyard gardener insecticide.

Bees die from direct contact of pyrethrum, but the material has such a short residue life it can be applied whenever bees are not actively foraging.

Microbial Insecticides

Microbial insecticides are developed from insect pathogens. Although thousands of insect pathogens occur among the viruses, bacteria, fungi, and protozoa, only a few have been developed into microbial insecticides. About 15 different insect pathogens are commercially available. Elcar, one of the nuclear polyhedrosis viruses, which infests corn earworms, has been the most extensively tested insect pathogen and the world's first naturally occurring virus to be registered as a pesticide. It is also the only one registered. Insect viruses, both polyhedrosis and granulosis types, seem to be completely harmless to both bee larvae and adults, at least the ones tested to date. However, bees have viruses like sacbrood attacking them; new insecticide viruses need to be tested on bees to ensure they don't harm them.

In 1980, the first protozoan, *Nosema locustae*, a

pathogen of grasshoppers was registered as an insecticide in the USA. This protozoan belongs to the same genus as the Nosema protozoan attacking honey bees. The grasshopper one doesn't harm bees and the bee one doesn't harm grasshoppers.

Recently, in Japan, the fermentation products of a bacterium yielded a remarkable new family of pesticides. Avermectin is an example. It's a broad spectrum material and kills differently than all other insecticides by inhibiting nerve transmission at the muscle. Avermectin is moderately to highly toxic to honey bees, but low in hazard to alfalfa leafcutting bees and alkali bees.

The bacterial insecticide *Bacillus thuringiensis* is a non-obligate pathogen easily grown on artificial diet. It forms a protein crystal that becomes toxic upon ingestion by insects. It also produces an exotoxin which kills insects. Many varieties of this bacteria are being developed for control of specific insect pests. Spores and crystals of the bacterium are harmless to bees, and the microbe does not cause bee paralysis during nectar flows. Tests with the exotoxin (DiBeta) of *B. thuringiensis* showed it is non-toxic to honey and alfalfa leafcutting bees. Tests with other exotoxins shortened honey bee life at extreme dosage rates.

PESTICIDES USED BY   7
BEEKEEPERS

Fungicides

The main fungus attacking honey bees is *Ascosphaera apis* while *Ascosphaera aggregata* attacks leafcutting bees. Both cause the disease known as chalkbrood. At present, no fungicides are available for control of chalkbrood though there may be some in the future. In feeding tests IPL-12, an alkyl amine, has shown promise as a chemical treatment to reduce chalkbrood of honey bees. Also, Benlate (benomyl) and Rovral (iprodione) show some promise for chalkbrood control. In experimental field tests, captan reduced the incidence of chalkbrood in leafcutting bees without harming bee larvae. However, in topical drop tests with honey bee larvae and feeding tests with leafcutting bee larvae, captan caused abnormal bee development and mortality. Pittclor (calcium hypochlorite) is registered in the U.S. to sanitize leafcutting bee cells, bee domiciles and nesting sites for chalkbrood control.

Herbicides

Many beekeepers use herbicides to control weeds

around the colonies. Roundup is probably one of the more commonly used herbicides around bees. Again, most herbicides are non-toxic to bees and can safely be used around the colonies.

Insecticides

Ants can be a major pest of honey bees. Ant nests found in the apiary can be destroyed with gasoline, diesel oil or an insecticide. Chlordan 5% dust, diazinon 2% dust or Dursban can be applied under the backs of hives. Bottom boards must be solid and hives off the ground so bees are not exposed to the insecticide. Commercial ant bait traps are available which can be placed under the hives or along ant trails. Do not use chlordane to control ants in bee equipment storage buildings. This insecticide has a strong affinity for beeswax, and its poison works for long periods of time.

Chlordane is restricted from use in many areas!

Wax moths are serious insect pests damaging combs and honey and killing honey bee larvae. Fumigants are often used to control this pest. The fumigant paradichlorobenzene (PDB) is effective on all but the egg stage of the wax moth and will not harm bees so long as the fumigated combs are allowed to air thoroughly before being placed in a hive. It also repels adult wax moths. Do not use PDB on combs containing honey intended for human consumption, as honey readily picks up the odor of the gas. For treatment, stack supers on a firm base and tape all seams and holes so the heavy gas doesn't escape through openings. If supers fit tightly, a tablespoon of PDB for each super is sufficient. For example for 6 supers, sprinkle 6 tablespoons (170 g) of PDB crystals on a piece of cardboard and place on the top bars of the uppermost super in each stack and put a tight-fitting cover over the top. Check the PDB crystals monthly and add more as needed. Damage to adult bees can occur from PDB so be sure and air all combs for 3 to 48 hours before reusing them in the colony.

Another fumigant is ethylene dibromide (EDB) which kills all stages of the wax moth, including eggs. Use EDB only on combs without honey or those with honey not intended for human consumption. EDB is a heavy, clear liquid which forms a colorless, nonflammable gas heavier than air. For treatment, prepare supers in the same way as to use PDB. Supers can be treated at the rate of 1 tablespoon (20 g) per eight-super stack. With the liquid, soak a small piece of felt building paper, burlap, or cloth and place it on the top super. Do not pour or discharge the fumigant directly onto the comb. Cover the stacks and seal cracks with masking tape. After 24 hours, remove the covering and aerate the supers completely.

In the U.S. EDB is no longer registered for this use.

Other insecticidal fumigants, such as calcium cyanide, methyl bromide, and naphthalene have been used to protect honey bee combs from wax moths. Ethylene oxide as a fumigant is only approved in a limited number of states and countries. This fumigant is effective against the wax moth and diseases such as American and European foulbrood and chalkbrood. All honey should be extracted from the combs before fumigation.

Certan (*Bacillus thuringiensis* Berliner) is a microbial insecticide for the control of wax moth larvae. The material does not harm bees. Each comb must be sprayed with or dipped in the insecticide solution for protection. Treatment will protect combs for up to 12 months.

Yellowjackets (wasps) can attack honey bee adults and larvae. When yellowjacket nests are found, use any one of the following insecticides. Pour or propel into entrance hole and do not plug the hole. Returning foragers will then enter the nest to be killed by the insecticide residue. Do after dark.

• Propoxur 1.5 EC, 8 ounces per gallon of water

• Carbaryl 5% dust poured into entrance hole without plugging hole

• Hornet and Wasp Killer (propoxur) is a good aerosol bomb.

Knox-Out 2FM (encapsulated diazinon) is registered to put into bait for control of scavenging yellowjackets. If Knox Out is used, the poisoned bait must be kept out of reach of children and pets.

Miticides

Miticides are used in many parts of the world to control the two major mites attacking honey bees: the honey bee tracheal mite (*Acarapis woodi*) and the varroa mite (*Varroa jacobsoni*). Hundreds of compounds have been tested as fumigants or systemics.

Tracheal Mite

Miticides used against tracheal mites include, in part, Amitraz (formamidine), "Mexican Acarol," Pentac (dienochlor), fluvalinate, Folbex VA (bromoprogylate) and menthol. In tests in Egypt, oil of clove, peppermint, marjoram, and the salt of eucalyptus controlled tracheal mites. Folbex is the main miticide used to control tracheal mites in India. Menthol, registered in 1989 in the U.S. shows great promise for tracheal mite control. Early experiments with menthol were done in Italy around 1965. Menthol packets containing about fifty grams of crystals and placed on the bottom board of hives will last 2-3 weeks. Although potentially lethal to bees in hot weather, menthol is economical and controls mites. Label approval for menthol has been granted from the Environmental Protection Agency for its use as a miticide. After 20 years of research, menthol appears to be a practical treatment for the tracheal mite.

7.1- Menthol crystals used for tracheal mite control

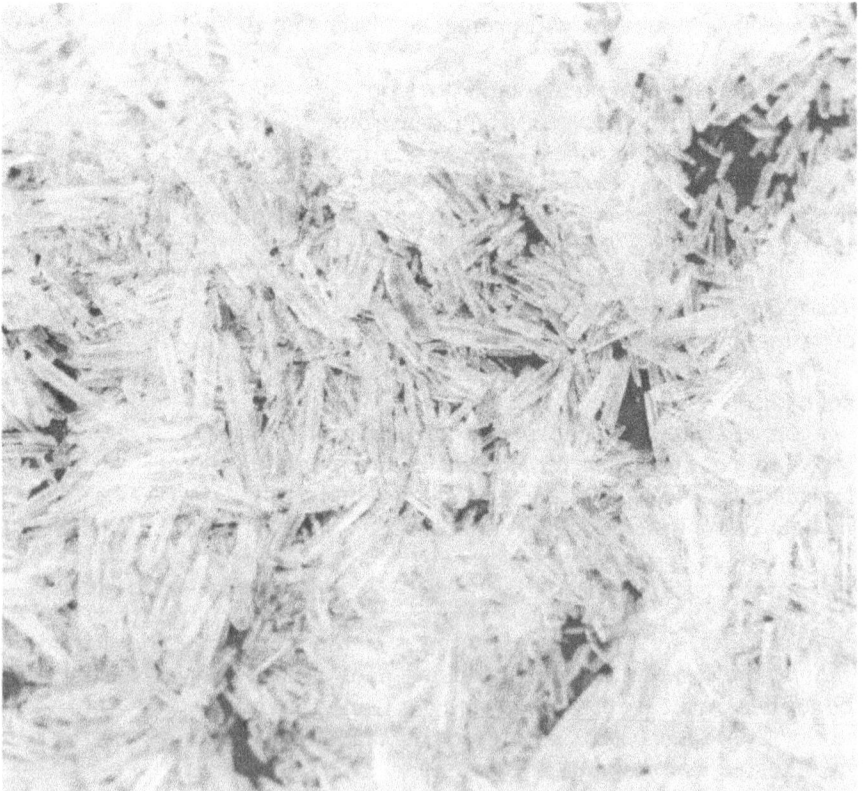

Varroa Mite

At present there are several chemical compounds under consideration for varroa mite control in the United States. There are materials registered in many other countries where the mite has been present for many years. Many pesticides including Dicofol, Kelthane, Morestan, nicotine, Omite, Pentac, Plictran, Propargite, and Tedion have been tested and used for mite control and are effective against varroa. Varroa mites in Japan are resistant to phenothiazine.

The following miticides appear to be the best materials for varroa mite control:

Folbex VA (bromopropylate) made by Ciba-Geigy is applied as a smoke by burning treated strips, either within the closed hive or in a device that injects the smoke into the hive. It is applied four times at four-day intervals to brood-free colonies. Treating bees with this product runs about $10.00 per colony and is labor intensive.

Another miticide is the Illertissen Mite Plate which is formic acid absorbed onto a special board and sealed in foil. The plates are placed on the top frames, the hive closed and the material fumigates throughout the hive. It can only be used early in the year or in the fall after the honey is removed from the colony and when there is no nectar flow.

Perezin (coumaphos) is a systemic compound that is applied to colonies by dripping it onto the bees. It is relatively safe for treatment of mite-infested bees, but can only be used 6 weeks before a nectar flow and not on small colonies.

Apitol is a systemic miticide containing formamidine compound and is fed to bees in sugar syrup or dripped onto the bees in a sugar carrier. It will probably not be registered in the U.S.

Only two miticides, amitraz and fluvalinate are now being considered for registration in the U.S.

Amitraz, made by Nor-Am Chemical Company, may be available in impregnated strips. It is also effective on the tracheal mite. However, recent reports from France indicate that varroa mites show resistance to amitraz.

Apistan (Mavrik, Spur) is a fluvalinate-impreg-

Amitraz

7.2 - Apistan strip for use in colonies for varroa mite control

plastic strips impregnated with fluvalinate

nated strip that beekeepers hang in the hive to kill varroa. It has also been used as an aerosol for varroa mite control in France. It is made by Zoecon Corp. and is used for mite control in many parts of the world. Fluvalinate is not toxic to mammals but is highly toxic to fish. Plastic strips impregnated with fluvalinate are placed in the brood nest. If they are placed on the tops of the frames, or on the bottom board, they are not as effective. Bees pick up the material on their feet and in their body hair as they walk across the strip and mites contacting the pesticide die. It works like a flea collar works on a dog. Fluvalinate plastic strips can be left in a colony for several months and will kill varroa mites as they emerge from the honey bee brood cells. Fluvalinate residues will probably not be a problem in honey, but they may be present in beeswax. Fluvalinate was registered for use against varroa mites in the United States during 1988.

WARNING: The above information is included for your information only, and should not be used as a recommendation. It is the beekeepers responsibility to check with local state, provincial and/or federal registrations for the above-mentioned products. Follow the label!

FACTORS CONTRIBUTING TO BEE POISONING

8

Bloom

Do Not Contaminate Bloom with Insecticides Hazardous to Bees

The contamination of open flowers (bloom) with insecticides is the major cause of bee poisoning. No wide-scale bee poisoning catastrophies have occurred without the insecticidal contamination of blooming plants.

Nearly 100% of all bee kills are directly linked to the application or mis-application of a hazardous insecticide onto bloom. Insecticide applications to crops or plants without bloom don't present a problem since bees won't be there. Bees pick up the insecticide when foraging on flowers or from other parts of the contaminated plant (figure 8.1). We found a difference in bee kills up to 27-fold in the presence of attractive blooms. Sometimes the foraging bees don't die, but they may pick up sub-lethal dosages which affect their behavior and reduce colony vigor.

All open flowers must be considered when we discuss 'bloom'. It may include flowers on the target crop

8.1 - While the 'target' crop is not in bloom, "weed" bloom poses a hazard.

8.2 - High density flowers, like clover, pose a larger hazard to bees than low density plants, like cucumbers.

(like apple flowers), flowers growing in the area of the target crop (usually called weeds, like dandelion, but may include clover and other flowering plants), or flowers outside the spray target area, where drift or misapplication may reach the flowers (clover blooming next to a sweet corn field could easily be contaminated unintentionally). Any open bloom poses a potential hazard to pollinating insects.

The amount of bloom in a crop often governs the number of bees visiting that crop(figure 8.2). The number of foragers affects the magnitude of a bee kill. For example, a highly toxic insecticide sprayed on an alfalfa field at 10% bloom results in less bee kill than the same material applied when the field is at 100% bloom — **more bloom, more bees, more kill**.

Usually, in urban areas, not all homeowners spray insecticides on the same day. Thus only a few blooming plants may become contaminated on a single day. Bees will die, but few symptoms of bee poisoning will be observed. On the other hand, a pest control company (tree or lawn service) may treat a several block area during a single day — contaminating a lot of bloom (figure 8.3). In urban areas a few colonies usually die

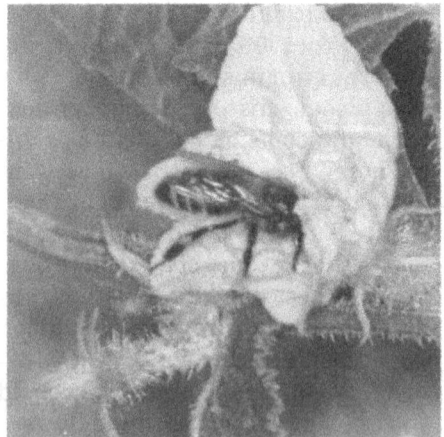

each year from foraging on contaminated plants when this type of spraying occurs. This has been reported in the Eastern United States where gypsy moth caterpillars defoliate most trees, and homeowners, homeowner associations, or town governments contract for the neighborhood treatment of all affected trees. Since these applications are usually made in May or early June, and use compounds which allow the colonies to build up in strength over the season, most colonies survive, but there is rarely any surplus honey.

The role of field size is clearly seen in spraying of large blocks of any agricultural crop. A 200 acre apple orchard sprayed during full bloom with a hazardous insecticide results in more loss of bees than spraying a 2 acre block in full bloom. Again, more bees, more kill.

In most cases, flower morphology or structure has little effect on the toxicity of pesticides to bees. Nor does flower color appear to influence bee poisoning. For example, the bee hazard of Spur (fluvalinate) on blooming carrot (an open umbrella type flower) is the same as on alfalfa (a closed type flower). There are a few exceptions, for example, when Thiodan (endosulfan) was applied to flowering broad beans and white clover, no bee mortality was observed Yet heavy bee mortality occurred when the same material was sprayed on cooumoellier — a much more open type flower.

Bees pick up the insecticide while foraging on the flowers. There is no difference between bee losses of colonies which have insecticides applied directly over the top of them and those sitting adjacent and not sprayed directly. It is the contamination of bloom — not beehives, which kills bees (figure 8.4).

Bees forage for honeydew (aphid secretions) on non-blooming crops, especially conifer trees and less frequently on corn. In North America reports of honeydew honey are less common than in Europe, and we aren't aware of any bee poisoning problems due to applications

◆
Do NOT apply nor allow pesticides hazardous to bees on blooming plants or crops.

8.3 - Residential lawn pest services will not cause bee kills unless they treat a large area of blooming plants.

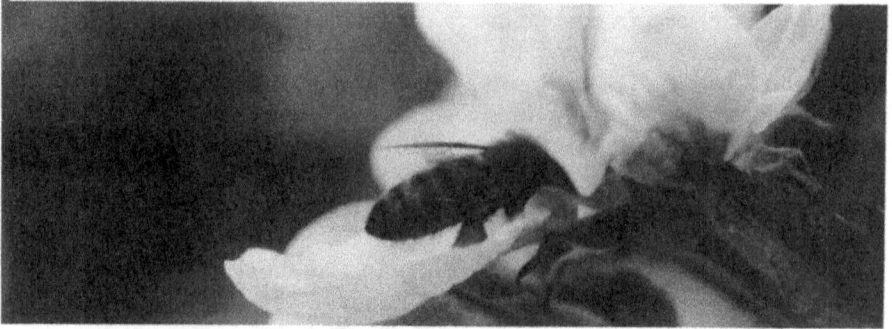

8.4 - Contaminated flower parts, not bee hives, cause bee kills.

to crops where honeydew honey is being produced.

Residue Exposure

Do Not Contaminate Bloom with Insecticide Residues Harmful to Bees

Residue is the amount of insecticide present on a plant after it's been sprayed. The amount of residue and its toxicity decreases with time as the chemical degrades. Residual action is of paramount concern because it largely determines whether an insecticide can be safely used on a blooming crop. For example, a material such as Dibrom (naled) can be applied with relative safety in late evening because of its short residual toxic effect on honey bees, even though the initial

RT_{25} hazard at application time is high. RT_{25} indicates the residual time (RT) required to bring bee mortality down to 25% in cage test exposure to field-weathered spray deposits. Some RT_{25} values are shown in figure 8.5 for common insecticides — a complete summary may be

8.5 - RT_{25} values of selected insecticides.

found in Appendix VI. Materials with an RT_{25} of 2 hours or less can be applied with minimal hazard to bees when bees aren't actively foraging. Those reaching RT_{25} within 8 hours present a minimal problem to bees, if they are applied during late evening or night. Those with RT_{25}'s greater than 8 hours cannot be safely used when they might contaminate bee forage. Variable RT_{25}'s for the same chemical are usually associated with low temperatures, since chemicals applied during cool weather retain a longer residual hazard. RT data

Parathion	13-18 hrs
malathion	6 hrs
methomyl	2 hrs
carbofuran	7 hrs
chlordane	<2 hrs
carbaryl, wp	7 days
naled	12-20 hrs
phosmet	>3 days
pyrethrum	<2 hrs

derived from field bioassays of insecticides for bees are given in Appendix VI.

Systemic insecticides may vary in bee hazard with age of foliage. The RT_{25} for methamidophos is less than one day for the leafcutting bee when treating young, succulent alfalfa; while the residual hazard increases to greater than 5 days when treating old, senescent foliage.

◆

Residual hazard of insecticides determines effects on bees.
(See Appendixes II & III)

Air Temperature

Change Your Insecticide Applications in Relation to Weather

Temperature affects residual action of insecticides. In general, there is a dramatic increase in residual killing action of insecticides occurring with low temperatures. **Chemicals applied during cool weather retain a longer residual hazard** (figure 8.6).

The residue hazard of Furadan (carbofuran) varies from one week to greater than two weeks when temperatures are cooler. DDT and Sevin (carbaryl) are considerably more toxic to honey bees at low than at high temperatures. For example, the bee toxicity of DDT decreases by a factor of 4 for every 10° C. rise in temperature between 15 and 34° C. Also, low night temperatures greatly increase the residual toxicity of Phosdrin (mevinphos) to bees.

8.6 - Low temperatures often increase residual hazard.

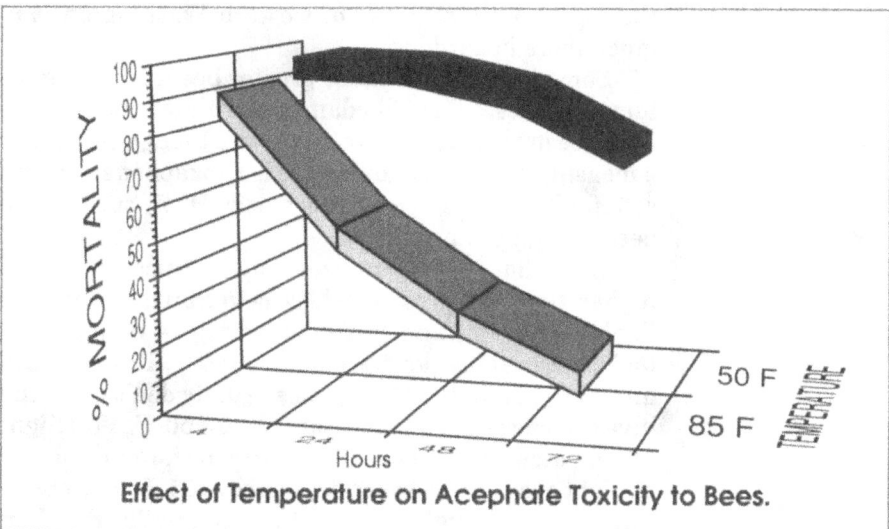

Effect of Temperature on Acephate Toxicity to Bees.

◆

Cooler night temperatures increase bee hazard of most insecticides.

Thirteen-day residues of Lorsban (chlorpyrifos) held at 50°F. caused greater levels of mortality than 7 day residues held in variable day-night temperatures, approximately a 2-fold longer residual toxicity. Orthene (acephate) residues held at 50°F gave an 18-fold longer residual toxicity than residues held at 64-95° F. Eight-hour residues of Pydrin (fenvalerate) held at 50° F. give a 2-fold higher residual toxicity than residues held at 64-95° F. Spur (fluvalinate), an insecticide considered non-hazardous to bees, gave a 30% increase in honey bee mortality at cooler temperatures. Unusually cold nights following hot summer days cause condensation of copious dew on the foliage, presenting a bee poisoning problem. Under these conditions, the residual action of insecticides increase and many more bees will be killed the following day.

Regional differences in the hazard of a given pesticide can often be explained in terms of differences in climate. For example, malathion often has a fumigant effect on bees in warm California, but does not in cooler Washington. Phosdrin (mevinphos) normally has a short residual effect in California and can be used when bees are not flying. Conversely, it sometimes continues to cause significant mortalities through one full day in Washington and cannot be safely applied to blooming crops. Also, in arid eastern Washington, Lannate (methomyl) is safe for bees if applied in late evening or early morning before bees forage. In the less arid midwest and the high humidity areas of western Washington it is much more hazardous.

Some insecticides cause greater bee kills at higher temperatures. For Thiodan (endosulfan) there is an increase in toxicity with temperature though this is not a measure of residual toxicity. Also, toxaphene appears safe for bees at temperatures below 64° F, but toxic to bees at high temperatures.

Immediate effects on bees may be much greater at higher temperatures — whereas residual effects are likely to be less, and less long-lasting — because the toxic material breaks down more quickly. Temperature affects bee activity. Honey bees usually don't leave the hive to forage with temperatures below 50° F. Full flight doesn't occur until above 55° F. Also, no foraging occurs at night even with temperatures above 55° F. A problem can occur with night temperatures above 70° F., when bees cluster on the outside of the hive. Bees in crowded hives cluster more than those in uncrowded hives. Clusters hanging at the hive entrance are like ripe grapes if an insecticide drifts onto them. They die, whereas bees in the hive will survive.

Timing

Do Not Apply Harmful Insecticides when Bees are Actively Foraging

The time of day of an insecticide application directly impacts on the hazard to foraging bees. **No chemical harmful to bees should ever be applied to bloom during the day when bees are foraging.** Proper timing during late evening, night, or early morning provides relative safety to bees from short-residual chemicals (figure 8.7). The residual killing power of the insecticide determines when it can be used. We've aimed much of our research at determining which insecticides can be used during off-bee hours without harming bees. Appendixes II through V list the insecticides stating which can never be used on bloom and the time of day when others can be used without harming bees and those that can be applied at any time without hurting bees. The tables are based on extensive toxicity and timing tests.

Bee-attractive plants requiring spraying while in bloom should be treated at night or in the early morning or late evening when the bees are not flying.

Geographical region affects safe timing. In the Pacific Northwest, the period between about 6 pm and 7 am is usually the safe non-foraging period for honey bees. In California, bees commonly forage between 4 am and 8:30 pm whenever temperatures rise above 60° F. In tropical areas with higher temperatures, bees are likely to forage whenever there is enough light to see. The safe daylight period for applying insecticides hardly exists. Lannate (methomyl) has a high initial hazard to bees, but a short residual toxic

8.7 - Don't spray when bees are foraging on the target crop or anything in the area!

71

effect on honey bees and can be applied in the evening without harming bees. On the other hand, a material such as Lorsban (chlorpyrifos) has a high initial hazard and remains hazardous for a number of days.

Bees forage on most crops all day as long as there is bloom in the field. In these cases the general rule of applying insecticides only in the evening or early morn-

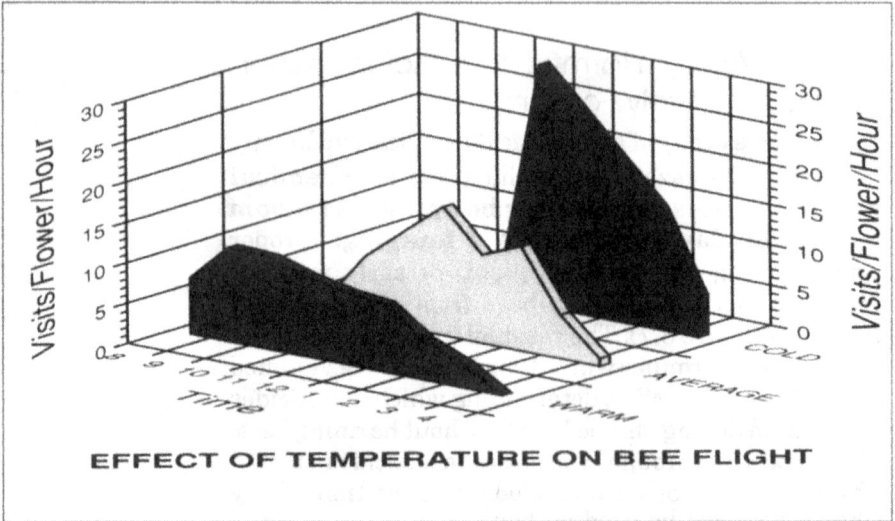

EFFECT OF TEMPERATURE ON BEE FLIGHT

8.8 - Comparison of bee flights on days with cold, average and warm temperatures

ing applies. Other crops like cucumbers and melons usually attract bees only from mid-morning through early afternoon. However, on warm mornings, flight may start as soon as there is enough light for the bees to see (figure 8.8). This provides less time for applying insecticides when bees aren't foraging.

Honey bees collect pollen from corn (and milo) even though corn is wind pollinated. In the western United States they have a specific time for foraging on corn, typically between 8 am and 1 pm each day. We found that 85% of the corn anthers are exserted from the male flowers between 4 am and noon, following the lowest temperature and highest humidity period. As temperatures rise, the pores at the bottom of the double anther tubes open and the pollen begins to sift out. Bees enter the corn fields and actively collect the pollen as it begins to be shed. Thus, short-residual insecticides can be applied to corn between 1 pm and midnight with minimal hazard to bees. However, this timing shifts with cooler temperatures of late summer and fall when the

Table 1. *The general rule for timing pesticide applications in such a way as to prevent hazard to bee pollinators.*	
HAZARD LEVEL	**TIME OF DAY**
Safest	late evening/night (from when bees cease foraging to midnight)
Intermediate	midnight to first light
Dangerous	early morning before bees begin foraging

foraging period can be 11 am till 4 pm or later.

In Wisconsin, on sweet corn, bees forage for dehiscing pollen from sunup to sundown for a period of 14-16 days. Insecticides harmful to bees can only be applied after sundown and before sun up.

Kills of foraging bees are often 2-4 times greater when applications are made in early morning as when they are made in late evening. Night applications, with floodlights mounted on wings of airplanes or on tractors, are being used on an increasing scale, especially where critical bee-pollinated crops are involved. In addition, some growers are finding that they obtain better pest control from such night-time applications and may not need to use as much material.

Many states and countries have regulations stating the time of day when certain pesticides can be applied.

Check with your state or provincial department or ministry responsible for pesticides to find out what the rules are concerning potentially hazardous sprays to bees.

Occasionally, the time of season is important, namely, treating when optimum control of a crop pest can be achieved and be least harmful to bees. An example is pre-bloom insecticide application on alfalfa seed for control of lygus bugs.

◆

Check residue hazard of pesticides before using on bloom or where it may drift onto bloom.

Formulation

Use a Low Hazard Formulation

Most pesticides are not pure active ingredient, but are mixed with other materials to aid in their safety, effectiveness or ease of application. Often the active ingredient or part that actually kills insects is only a small fraction of the packaged product. Insecticides are available as dry or liquid formulations.

The most commonly used dry formulations are dusts, granules, wettable powders, and soluble powders. The most popular liquid formulations are emulsifiable concentrates, oils, solutions, flowables, fumigants, and aerosols. Some pesticides which are toxic to bees in one formulation may be much less toxic in a different formulation.

Chemical companies make **dust (D) formulations** by grinding the pesticide into a fine powder. Then the pesticide is mixed with a carrier such as bentonite, diatomacious earth, pyrophyllite or talc. Average particle size ranges from 2 to 75 microns. The use and manufacture of dust formulations has dropped off in recent years. It is an expensive way to formulate pesticides and it is especially susceptible to drift.

Companies make **wettable powders (WP)** in a similar manner — by grinding the pesticide into a fine powder and then adding a wetting agent and a sticker. Wetting agents make the material mix well with water. Stickers such as drying oils or casein or other adhesive agents make the residue cling to the sprayed surface of the plant.

Soluble powders (SP) are powder formulations that dissolve in water and do not require wetting agents or agitation.

Flowables (F) are similar to wettable powders except that the finely ground insecticide is suspended in water and surfactants (detergent-like compounds) are added to make the insecticide solubilize in water.

Solutions (S) are stable formulations of active ingredient dissolved in a compatible solvent without an emulsifer.

Emulsifiable concentrates (EC) are concentrated oil solutions of the pesticide to which an emulsifying agent has been added. The concentrate, when added to water, forms a suspended emulsion of minute oil droplets.

Aerosols are a pressurized solution of pesticide in a propellent mixture. Aerosols for flying insects have particle size of about 10 microns. Other aerosols are designed to deposit residues on surfaces.

Granules (G) are similar to dusts except the particle size is much larger — in the range of 1 to 2 mm. Granules are usually applied into the soil or broadcast on the surface of the ground. They are seldom used on blooming plants, and are essentially non-hazardous to bees. The only exception may be systemic insecticide granules applied before bloom.

Formulation affects bee hazard of insecticides. Dust formulations are usually more hazardous to bees than sprays. Wettable powders often have a longer residual effect than emulsifiable concentrates.

The typical sequence of bee hazard of different formulations of insecticides (from most hazardous to least hazardous):
- **dust D**
- **wettable powder WP**
- **flowable F**
- **emulsifiable concentrate EC**
- **soluble powder SP**
- **solution S**
- **granular G**

This difference in bee toxicity of different formulations is related to pick-up of the toxicant.
- For example, Lannate (methomyl) dust is in the most toxic group while the liquid and soluble powder formulations are in the less toxic group.
- In comparative laboratory and field tests with five formulations of Sevin (carbaryl) a slight alteration in physical properties significantly reduced the hazard to honey bees. The finer the grind of Sevin, the safer the material. Adding a sticker to the formulation made Sevin 4 to 5 times safer to honey bees.
- In comparative wettable powder vs. liquid formulations tests, the wettable powder formulations caused as much as a 6-fold greater kill of honey bees. This occurred even when treated foliage contained more of the active ingredient following liquid formulation applications.

◆

For pesticides with different formulations, choose the formulation safest to bees.

A process called **'pick-up'** is a major factor in the surface contact action of insecticides against honey bees. Pick-up may be associated with the branched or otherwise modified body hairs of bees, which are adapted for holding minute pollen grains. Differential sorption of the pesticide into plant tissues doesn't appear to be a major factor in pick-up. Analysis of insecticide residues in surface washings, wax strippings, and tissue extracts of foliage treated with various formulations showed no significant differences in profiles of sorption.

A recent innovation is **microencapsulated insecticides**. Micro-encapsulation of a pesticide extends the effective life of a short-lived pesticide by releasing it slowly through the plastic walls of the capsules. The insecticide is incorporated into polymeric microcapsules that are about the same size (30-50μ) as pollen grains. Microencapsulated insecticide formulations appear to be extremely hazardous to bees because electrostatic charge and capsule size are similar to pollen (figure 8.9). This results in a marked increase in their pick-up by bees. These capsules adhere readily to the bees and are

8.9 - Microencapsulated insecticides are picked up by the branched hairs on the bee's body.

ultimately brushed and combed into the pollen pellets on the corbiculae (pollen baskets) of the honey bee's legs. When contaminated pollen is stored in the frames it can lead to colony decline by killing bees which eat poisoned pollen. Or, severe bee poisoning problems involving kill of bee brood and newly-emerged workers may occur.

One microencapsulated formulation of methyl parathion (Penncap-M) retains a toxic hazard to bees in

stored pollen from one season to the next, and causes a delayed break in the brood cycle, similar to that experienced from carbaryl dust formulations applied to corn. Our field studies show Penncap-M causes significantly higher residual bee mortality than methyl parathion EC. In a Connecticut study, colonies exposed to Penncap-M were significantly more likely to die than colonies exposed to any other material.

Colony Strength

Strong Colonies Suffer Higher Losses

Strength of the honey bee colony definitely affects how many bees die, though it doesn't affect toxicity of the chemical. Populous colonies always suffer greater losses than weak colonies because greater numbers of foragers are exposed to the insecticidal residues. Often the kill is at least four times as great with the same application as compared to weaker colonies. However, strong colonies usually bounce back from a bee loss better than weak colonies.

In cases where heavy kills of foragers occur frequently (weekly), far few foragers leave the colony. This is due to the fact that most of the remaining bees stay in the hive to cover and tend brood. Until new bees emerge, few or no bees forage and the colony suffers no additional loss of adult bees. In one study, some of the weakest colonies didn't lose any adults to insecticides for 14 days simply because no bees foraged.

◆

Strong colonies often suffer greater losses, yet they may recover from pesticide losses faster than weaker colonies.

Distance

Distance of Bees from Treated Fields Affects Bee Losses

Distance of colonies from treated fields is inversely proportional to the amount of mortality that occurs. A certain amount of distance is required for this effect to be evident. For example, we see no difference in the amount of bee kill between colonies set in the treated field versus those within 100 yards of the field. On the other hand, we've noted up to a 9-fold reduction in kill as little as a half mile from the treated crop (figure 8.10). In general, in diversified agricultural areas with lots of different blooms, injury isn't significant to colonies a quarter mile or more away from the treatment.

Conversely, during a dearth of available pollen and nectar, bees may be severely poisoned by treatments applied at a considerable distance from the apiary. This can happen when the treated crop is the only attractive field in the area. Such bee kills have occurred at distances up to three to four miles.

The farther the colonies are from the treated area,

◆

In diversified areas try to set apiaries more than a quarter mile away from crops that may be sprayed with insecticides while in bloom; set holding yards 4 miles from orchards.

the less critical the time of day of the treatment.

Colonies moved into a treated field 2 or 3 days after treatment usually escape damage. But, if colonies are moved in too early, they may receive even worse damage. Colonies left adjacent to a treated field may suffer less damage than colonies moved next to the field several hours after a morning application. Bees from the moved colonies don't fly far from home and all foragers will concentrate on the treated field.

Forage

Lack of Alternate Forage Aggravates Bee Poisoning

8.10 — Colonies positioned away from a treated field or orchard experience less pesticide damage.

Lack of suitable alternate pollen and nectar plants severely aggravates bee poisoning problems in many areas. Ever-increasing monocultures of crops, and the removal of fence-rows, wild strips and wasteland reduces bee forage and nest sites for wild bees. Since the late 1940's these developments have forced beekeepers to pasture their bees on insecticide-treated fields. Development of efficient herbicides results in severe reduction in bee forage on both cultivated and wild lands. This involves not only direct removal, such as elimination of sweet clover from wheat fields, but also treatments which cause displacement of bee plants by herbicide-tolerant plants yielding nothing to bees, such as Dal-

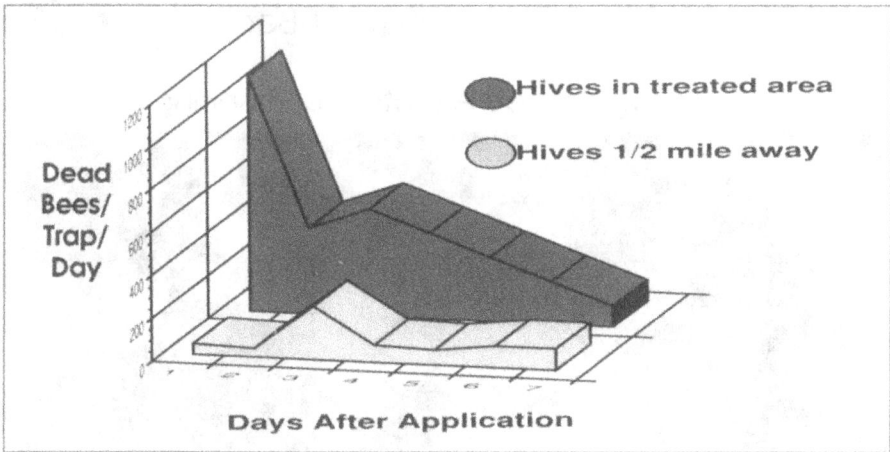

Dead Bees/ Trap/ Day

Days After Application

Hives in treated area

Hives 1/2 mile away

matisan toadflax, yarrow, and nightshade. There can also be a conflict of interests when noxious weeds (yellow starthistle, hound's tongue, Canada thistle) are excellent pollen and nectar sources.

Pollen availability on other crops can reduce the bee poisoning problem on corn. Fewer bees will be working the treated corn. Several management practices, including feeding pollen-supplement cakes and water reduce the effects of poisoning. Provision of pollen cakes helped prevent bee mortality in Wisconsin, but the use of pollen traps on hives didn't reduce bee mortality from gypsy moth sprays in Michigan.

We tried developing bee forage preserves to counteract this trend, but found it not feasible. We planted good forage plants on land never intended for farming. The procedure we used is discussed in bee forage preserves (Chapter 15). Another approach to lack of bee forage involves land-based honey production. That is: the plant's only commercial value is as a nectar source to honey bees. In a 1979 survey of 1,466 Washington beekeepers, lack of bee pasture was perceived to be a very serious problem. Pesticide related bee kills were considered to be the most serious problem. In many respects, forage and bee poisoning are inter-related problems. Land-based beekeeping requires the beekeeper to become a farmer and buy or lease land to raise the nectar crop. In this type of beekeeping the farmer/beekeeper has direct control over the forage plants and pesticide use on those plants.

8.11 The closer colonies are to the treatment site, the greater the bee loss

◆

Find or plant uncontaminated forage for your bees.

Age of Bees

Age of Bee Affects Bee Poisoning

8.12 - Smaller insects require less chemical to kill them than do larger insects. Shown here are a leafcutting bee, honey bee and bumble bee.

◆

Keep colonies rearing brood for a good mix of age of bees.

Age of honey bees affects their tolerance to insecticides depending upon the specific compound; newly emerged bees are most susceptible to DDT, dieldrin, and Sevin (carbaryl), and older bees are most susceptible to malathion and methyl parathion. The greater brain acetylchoinesterase (AChE) concentration in young honey bees gives them their greater tolerance to malathion.

After leafcutting bee females actively nest in the field for 3 or more weeks, they become much more susceptible to insecticides. We've shown this with Dylox (trichlorfon), Dibrom (naled), Carzol (formetanate), and Lannate (methomyl) and, not surprisingly, the untreated checks also increase in mortality with age of bee.

Body Size

The Species of Bee and Body Size Affects Bee Poisoning

Body size appears to have a direct effect on the susceptibility of bees to insecticides. As a general rule, larger bees are more tolerant of insecticides than are smaller bees. Smaller bees have a higher surface-to-volume ratio and are more susceptible. For example, endrin has less than 2 hours residual toxicity to honey bees, more than 3 hours to alkali bees, and over 24 hours to alfalfa leafcutting bees. The leafcutting bee tends to be more susceptible than the honey bee to many insecticides.

Tropical African and Africanized honey bees have a slightly higher surface-to-volume ratio than European bees. Therefore, they might be more susceptible to insecticides. Using the same ratio, we would suspect insecticides would be more toxic to *Apis cerana* and *Apis florea* and less toxic to *Apis dorsata*.

In addition, AChE specificity differs between species and reductions of this enzyme correlate with development of signs of organophosphorus insecticide poisoning.

Many factors affect the hazard of pesticides to bees. In addition, alfalfa leafcutting bees, alkali bees, bumble bees and honey bees often react differently following exposure to pesticides. Alfalfa leafcutting bees are inherently more tolerant to insecticides than alkali or honey bees as shown by topical drop (LD_{50}) and acute oral studies. Alkali bees are intermediate in inherent tolerance to insecticides. Enzyme systems were examined as possible sources of variation in bee species tolerance. The only positive correlation found was the higher pH of the body fluid in the leafcutting bee.

In what appears to be an anomaly, the inherent insecticide tolerance of the leafcutting bee reverses under field conditions. Since essentially all cases of poisoning occur from chance adherence of insecticidal residues to bees foraging treated plants, we suspected the main reason for susceptibility in the leafcutting bee was its greater surface-to-volume ratio. We made detailed measurements and weighed fresh samples of alfalfa leafcutting, alkali, and honey bees. Analyses provided the following data for the leafcutting bee female, the alkali bee female, and the honey bee worker. The surface-to-volume ratios (mm^2/mg) were: leafcutting bee, 94/33; alkali bee, 165/87, and the honey bee, 186/128. The index for leafcutting bee, alkali bee and honey bee was 2.0, 1.3, and 1.0 respectively.

◆

Field suscepti-
bility sequence:
most suseptible
alfalfa leaf
cutting bee
> alkali bee
> honey bee
> bumble bee,
least
susceptibe

Selectivity

Some Insecticides Do and Some Don't Harm Bees

The ideal selective material doesn't harm bees, yet kills pest insects. Some of these exist and combined with timing of application form a major part of a poisoning prevention program. Appendixes II through VI list the relative hazard of many pesticides.

Two factors govern susceptibility of bees to any insecticide:

- selectivity or toxicity
- formulation

Both can be manipulated to reduce bee losses.

◆ **Use selective, non-hazardous to bees insecticides. (see Appendixes II through VI)**

Hazardous insecticides stress colonies. And, the amount of other stress the colony is under affects, to some extent, damage inflicted by the insecticide on the colony. Strong colonies under limited stress survive bee poisoning better than colonies under high stress. Sometimes, the effects of an insecticide on a colony of bees differs under different environmental conditions. Therefore, so does the selectivity of the material.

Pesticide Application Method

◆ **Keep the chemical on the target.**

Use Low Hazard Method of Application

Pesticides are applied by fixed-wing aircraft, helicopters, through the sprinkler lines or by ground equipment. Airblast sprayers, foggers, and backpack sprayers are several ground application methods. **Potential risk to bees is highest with aerial application.**

Drift or the movement of the pesticide off the target area causes most bee kills. Aircraft applications present maximum bee hazard because of the potential for drift. For example, in 1984, in Washington, of 24 reported bee kills, 20 resulted from aircraft applications and drift. Drift can also be a problem with airblast sprayers. Insecticides should never be applied when the wind blows and it's impossible to keep the chemical on the target crop.

The outcome of drift can be catastrophic. Small pesticide particles in the air blown by the wind onto blooming crops or weeds is a major factor in bee poisoning. Often, pesticides applied to crops not attractive to bees present problems as the wind blows the material onto cultivated crops or to nearby weeds where bees actively forage for nectar or pollen. Pesticide drift most damaging to bees usually originates in an adjacent field or one a short distance away. However, drifts of insecticides over several miles have resulted in serious adult bee mortality. For example, in 1988 an ULV malathion application to cherry drifted 1.5 miles onto a blooming alfalfa seed field killing more than 80% of the bees.

Insecticides shanked into the soil or granular formulations broadcast on the ground present minimal bee hazard.

In general, with applications confined to the target, coarse sprays are more hazardous then fine sprays or aerosols. The potential for drift of dusts, fine sprays and aerosols exceeds that of coarse sprays.

Myths About Bee Poisoning

Understand Ways Pesticides Do NOT Kill Bees!

Myth? Aerial applications should not be made over honey bee colonies
If you leave a sizable area — more than several acres — around an apiary, it will help protect the bees from poisoning. However, if groups of honey bees are placed within or adjacent to a field to be treated, refraining from spraying over the hives doesn't seem to change bee mortality. We have observed the same results whether hive entrances were covered during spraying or not and whether a swath over groups of hives was left untreated or not. No doubt it is good practice to leave untreated swaths, just to be on the safe side.

Myth? Bees can be killed by simply flying over a treated field
Beekeepers sometimes claim that their honey bees were killed by flying over treated fields. However, there is no documentation to show that this occurs. An easier explanation may be seen via this example:
In one case, the beekeeper noted that his bees were

flying over a parathion-treated field of wheat (where bees do not forage) and not returning, except for a few which were dying at the hive entrance. Investigation showed that the bees were getting poisoned on patches of blooming Canada thistles (*Cirsium arvense*) out in the middle of the field.

In California, E.L. Atkins investigated this as well, and found bees were not killed when they flew over treated fields.

Myth? Colonies of honey bees placed in a low area can be poisoned by insecticide vapors which move from treated fields nearby and settle in the hive area

Neither E. L. Atkins nor we have observed bee kills which could be attributed to movement of vapors from a treated field. We did not get fumigating action from organophosphorus insecticides such as dimethoate, dibrom, or diazinon when we placed cages of bees in treated plots immediately after application. However, E.L. Atkins did obtain a fumigant action from such materials following treatment. Clearly, such differences must be related to climate and weather.

Myth? Mixtures of more than one pesticide are a special hazard to bees

Sometimes applying mixtures of two or more insecticides reduces the hazard to bees as compared to applying them singly. When the insecticides are in emulsifiable concentrate formulations, we suspect the increased quantity of solvents per gallon of spray provides a safer effect. Also, sometimes we use two or more selective materials which are less hazardous than a single broad-spectrum material. One exception is the use of certain insecticide-specific miticide combinations. In such cases, the miticide appears to have a synergistic effect which causes the mixture to be more hazardous.

The additive effect of two or more chemicals — cannot be discounted entirely. In a survey of pesticide-exposed colonies in Connecticut, those colonies which showed residues from two to even five compounds were much more likely to die than colonies which were found to contain only one compound.

MORTALITY FACTORS CONFUSED WITH POISONING

9

Winter Kill

Starvation is a major cause of death for honey bees kept in cold climates. It typically causes losses of 5 to 10% of colonies, depending upon the winter severity in a given area. It can be easily distinguished from chemical poisoning because of different symptoms. Winter kill (starvation) causes the dying bees to remain on the top bars and head first into the empty cells, conforming to the shape and size of the winter cluster of bees. With chemical poisoning there are no furry patches of bees with their abdomens sticking out of the honeycomb cells and no bees encrusted on the top bars of the frames.

Poison Nectars and Pollens

Honey bees collect nectar and pollen from a great many different kinds of plants. A few plants (15 to 20) produce either nectar or pollen which is toxic to honey bees. A particularly interesting phase of plant poisoning is the extremely specific action of the toxic substances from different plants. In some cases only the field bees die. In other cases, losses have been limited largely to recently emerged bees, to sealed brood or to unsealed brood. In some cases, both the adult bees and brood

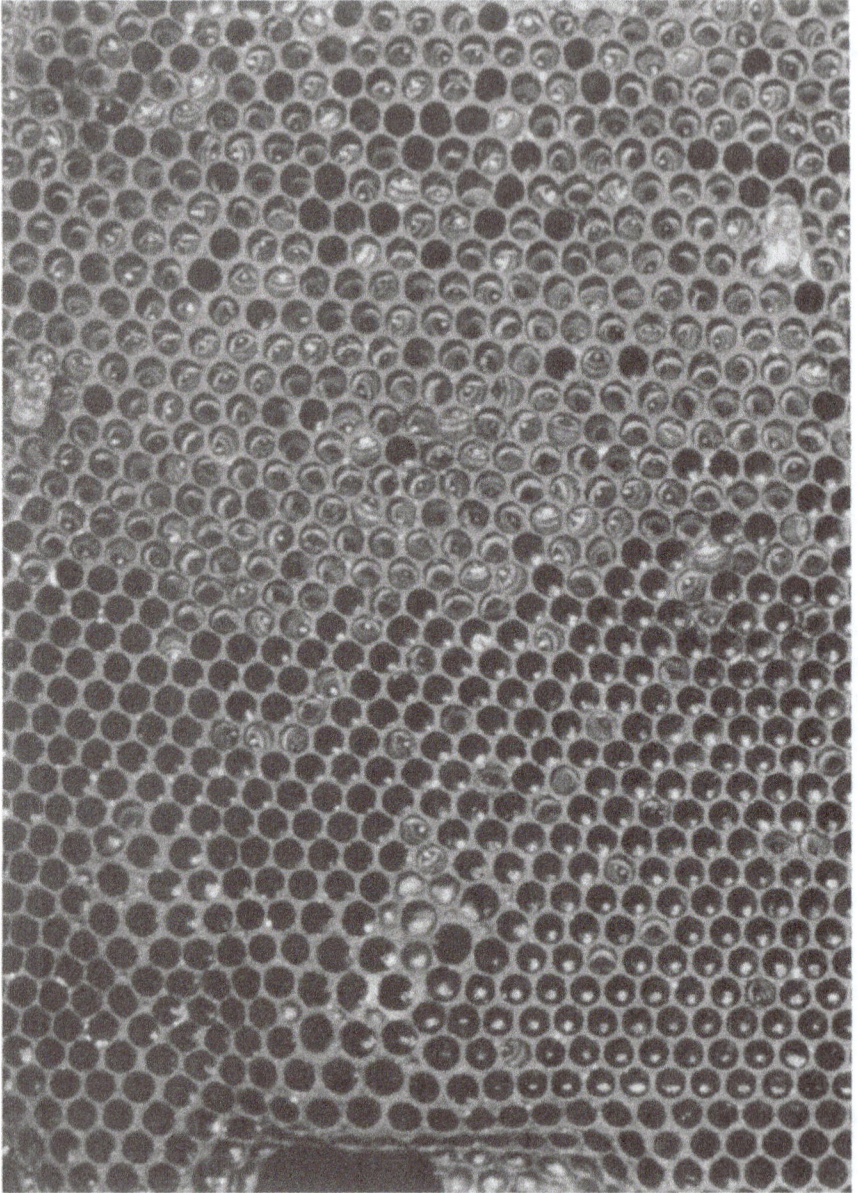

9.1 - When honey bees die of starvation, they are often inside the cell, head first.

appear to be affected by poisons from the same plant.

California buckeye (*Aesculus californica*) is one of the most important and best-documented poisonous plants. It causes a considerable foraging area to become unsuitable for bees during the early summer. Most other plants such as *Rhododendron* and *Astragalus* species, are less widely distributed and generally are of less concern. Usually, the development of a poisonous plant problem is associated with unusual drought or cold conditions in a certain area.

In general, honey from these plants have

9.2- California buckeye causes bee deaths

NOT harmed humans. Honey from the mountain laurel (*Kalmia latifolia*) may cause numbness or cause loss of consciousness for several hours though no other effects have been reported.

Symptoms of plant poisoning of bees vary widely, and it is not easy to make a positive diagnosis of plant poisoning by examining a sample. Neither chemical analysis nor microscopical diagnosis can be used to detect plant poisoning.

There are a few general characteristics, however, that appear to be more or less constant. Plant poisoning usually comes on suddenly, and after continuing for a more or less defined time, disappears. In most cases, most colonies in an apiary are affected at or about the same time and in a similiar degree. At times, however, there is considerable difference in the extent to which different colonies within the same apiary are affected. Some colonies may suffer no noticeable losses, while others become weakened or die. If the nectar is poisonous, symptoms occur only during the blooming period of the plant, if the colony survives, the symptoms disappear with the bloom. However, if the pollen is toxic, the symptoms may linger as long as the pollen remains in the combs.

Symptoms of plant poisoning in adult bees are somewhat similar to the symptoms of pesticide poisoning. Adult field bees die in and near the hive, in the field, and on the flowers of the poisonous plant. There may be piles of dead bees in front of the hive or not enough adult bees in the hive to take care of the brood. Poisoned bees often become weak, unable to fly, and crawl on the ground. New adult bees may have crumpled wings. In advanced cases of buckeye and loco-weed poisoning, adult bees frequently tremble and shake as in an advanced stage of paralysis.

Brood affected by plant poisons may die at anytime between hatching and adult emergence. In purple brood, considerable young bees die in sealed as well as in open cells. Dead brood generally lacks the brown or black colors associated with foulbrood. In buckeye poisoning, most of the brood dies shortly after hatching. Also, queens affected by buckeye poisoning lay fewer eggs, in an irregular, scattered pattern. Many of the eggs do not hatch.

With some symptoms it may be difficult to distinguish between plant and pesticidal poisoning. Usually plant poisoning is a annual problem in the same area year after year. Beekeepers become familiar with the poisonous plants and the location and time of the bloom. Also, if symptoms occur in a non-agricultural area where no insecticides are used, the symptoms are probably due to plant poisoning. In addition, in most agricultural areas, there are few poisonous plants since most of the land is devoted to farming. Therefore, symptoms observed in agricultural areas are probably due to pesticides. Of course, chemical analyses showing pesticide residues in dead bees or hive products confirms pesticide poisoning.

Swarming and Absconding

Swarming may be confused with poisoning, especially if only 1-2 colonies of honey bees are involved. Suddenly, the bulk of the field force is gone, but none of the typical symptoms of poisoning occur. There is no mass of dead and dying workers at the entrance, no regurgitation, no signs of intoxication, no unusual number of deaths amongst newly emerged workers, no break in the brood cycle, and the colony quickly recovers, with no continuing loss of numbers.

Absconding is less common and, for European strains of honey bees, is mainly associated with lack of forage. It may become more of a factor in the future, since absconding is associated with Africanized bees.

Smog and Pollution

Researchers and shown that smog and pollution are not a serious problem for honey bees. In fact, air pollution will destroy the flowers bees visit before it will destroy the bees.

9.3 - When a colony swarms, about 1/2 the bees are lost from the parent colony. Such colonies may appear weakened by insecticides.

Diseases

Honey bee diseases, especially American foulbrood (AFB), can be kept to a minimum with proper management, which includes both treatment with drugs and good apiary hygiene. AFB is detected by several symptoms, including a rotten odor of the dead larvae, and the sticky consistency of the dead larvae and pupae (which 'rope' with a match test). Honey bee larvae which die following chemical poisoning are mainly killed by desiccation. They simply dry up, and do not have the strong, rotten odor or the sticky properties of AFB victims.

Viral paralysis might possibly be confused with Sevin poisoning, because individual bees lose the ability to fly, crawl on the ground and foliage in front of the hive, and take several days to die after onset of these advanced symptoms. Thousands of dead and dying bees may accumulate in front of the hive, and they may become almost hairless, almost black, shiny or greasy in appearance and have bloated abdomens associated with dysentery. None of these latter symptoms are similiar to those caused by current insecticides (arsenicals did cause or lead to hairlessness).

9.4 - American foulbrood is a frequent cause of death in honey bee colonies. Sunken cappings, a rotten odor, and 'ropy' materials are tools for identification.

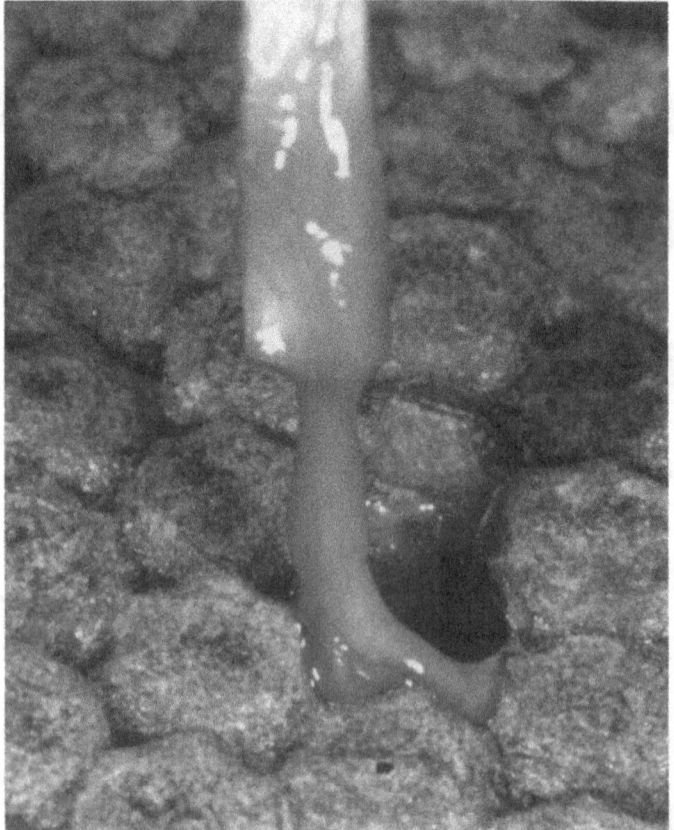

Parasites

Recently, two destructive Old World parasites of bees have invaded the United States and become widely distributed. One is the tracheal mite (*Acarapis woodi*) and the other is the varroa mite (*Varroa jacobsoni*). Researchers are actively working on control measures for both of these mites. At any rate, symptoms of these parasites should not be confused with the effects of pesticide poisoning. The slow debilitation of colonies and shortening of longevity of individual bees is contrasted with the short term and dramatic effects of poisoning.

Table 2. *Plants Known To Be Poisonous To Honey Bees*

mountain laurel	*Kalmia latifolia*
karaka tree	*Corynocarpus laeuigata*
California buckeye	*Aesculus californica*
black nightshade	*Solanum nigrum*
death camas	*Zygadenus venenosus*
dodder	*Cuscuta* spp.
leatherwood	*Cyrilla racemifloral*
locoweeds	*Astragalus* spp.
seaside arrowgrass	*Triglochin maritima*
whorled milkweed	*Asclepias subverticillata*
hellebore	*Veratrum* spp.
henbane	*Hyoscyamus niger*
horse chestnut	*Aesculus hippocastanum*
rhododendrons	*Rhododendron* spp.
yellow jessamine	*Gelsemium sempervirens*

10 FOOD CONTAMINATION

Nectar Contamination

Pesticides in Nectar are not a Major Problem

Nectar is produced by glands called nectaries which are usually located in the flower (figure 9.1). The glands secrete a solution of dissolved sugars or nectar which attracts bees so that the flowers may be cross-pollinated. Plant roots take up systemic insecticides from the soil and translocate the material throughout the plant. With some systemics the insecticide enters the plant through the foliage and transfers throughout the plant. Potential problems present themselves with nectar contamination by plant systemic insecticides. **Most systemic materials tested have not been hazardous to bees at standard field dosage rates.**

Only two insecticides manifest a nectar contamination hazard for bees. In greenhouse studies, Temik (aldicarb), at twice the recommended rates, caused moderate to high mortalities of alfalfa leafcutting and alkali bees for 14 days. In field studies, aldicarb granules injected into the soil at 3 lb of active ingredient (ai) per acre (recommended rate) and irrigated during bloom, reduced reproduction by alfalfa leafcutting bee females. Conditions of dosage, timing and irrigation required to eliminate this hazard are quite restrictive, so aldicarb is not recommended for use on alfalfa grown for seed where wild bees serve as primary pollinators. Temik isn't hazardous to bees under recommended field use conditions, since they are not exposed to contaminated nectar until at least 4 weeks after treatment. Temik repels honey bees from blooming carrot.

In greenhouse studies, Cygon (dimethoate) granules injected into the soil at 10 lb ai or more per acre kills alfalfa leafcutting bees through contamination of nectar for about 15 days. At the recommended field dosage on alfalfa, 0.5 lb ai per acre, it has no adverse effects. In sugar syrup feeding studies with honey bees, Cygon was hazardous at 0.2 ppm to small colonies in caged minihives. However, full-sized colonies fed either 0.1- or 1.0-ppm Cygon performed better than or as well as control colonies in production of sealed brood and maintenance of adult populations. In southwestern U.S. investigations, Cygon repelled honey bees from blooming lemon and onion, and was toxic to the bees.

Foraging honey bees have brought back to the colony nectar from alfalfa, birdsfoot trefoil, field beans, citrus, fuchsia, onion, and rape contaminated with Cygon (dimethoate). In addition, menazon, phosphamidon, Systox (demeton), Schradan and Thimet (phorate) have been found in nectar brought back to the hive by bees. Also, orthene (acephate) and Lannate (methomyl) have been found in nectar. But, again, using standard rates and under normal environmental conditions these haven't caused significant bee kills.

Contaminated nectar presents minimal, if any, hazard to alkali bee larvae because they eat essentially pure pollen. Freshly prepared pollen balls of alkali bee females only contain about 9% nectar. Many other wild bees have about the same amount of nectar in their rather dry pollen balls.

There are no instances of general contamination of honey by pesticides. Most pesticides are not in the nectar and bee behavior helps prevent contamination of the honey. Bees gathering poisoned nectar usually die in the field or in the hive before giving up their contaminated load. Sick foraging bees making it back to the hive demonstrate abnormal behavior and house bees remove them from the hive before they expel

◆

Most systemic insecticides don't harm bees.

their load of nectar. Guard bees also resist abnormal acting bees or bees which return with offensive odors and remove them from the hive. In bee poisoning investigations, detectable residues of pesticides have often been found in dead bees, pollen, and bee brood but not honey. In those instances where insecticide contamination has been detected in honey, the amounts found were considered too small to be significant. Also they were only found in honey cold extracted from frames and not filtered; it was not found in commercially processed honey.

Pollen Contamination

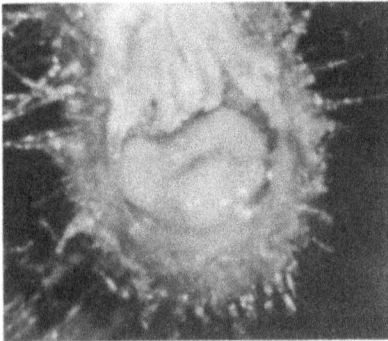

10.1 - Nectary of cucumber flower

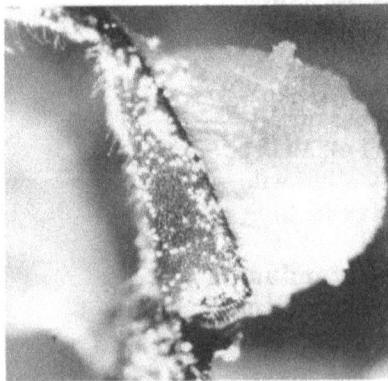

10.2 - Pollen pellet on hind leg of honey bee

Pesticides in Pollen is a Major Problem

Pollen contamination is the major source of the poisoning problems which affect entire honey bee colonies. Pollen-collecting workers moisten their appendages and brush and comb their body hairs to form pollen pellets on the corbiculae. Powder formulations of insecticides become readily incorporated into the pollen loads in this way. Microencapsulated insecticides pose a severe hazard because the plastic capsules are about the same size as pollen grains and have a strong affinity for adherence to the bee. Pollen contaminated with Systox (demeton) and nectar contaminated with Temik (aldicarb), but usually not vice versa, has been observed. Neither material is detrimental to leafcutting bees. Pollen-nectar stores of leafcutting bees in a certain area of Nevada contained unusually high toxaphene residues (up to 56 ppm). Toxaphene was definitely associated with increased levels of immature mortality in this case. Dylox (trichlorfon) and Dibrom (naled) have been implicated as causes of immature mortality of alfalfa leafcutting bee larvae in unreplicated field trials. Our tests showed these materials do not cause immature mortality under field conditions.

After field applications, Ambush (permethrin), DDT, ethyl parathion, Furadan (carbofuran), Lannate (methomyl), malathion, methyl parathion, Pounce (permethrin) and Sevin (carbaryl), have been found in stored pollen. Arsenicals remain in pollen stored in comb six months after application. Penncap-M (methyl parathion) has persisted in comb samples of stored pollen for 7 to 14 months after field applications. Pollen contaminated with insecticides has not been found in honey.

◆

Contaminated pollen brought into colonies can destroy honey bees.

Leaf Piece Contamination

Leaves Tainted with Insecticides May Kill Leafcutting Bees

Leafcutting bees use leaf pieces to construct their cells. Leaf pieces contaminated with an insecticide could be hazardous (figure 9.3). Leaves contaminated with DDT and Guthion (azinphosmethyl) killed leafcutting bees in experiments, but these materials are no longer used on alfalfa seed. Also, leafcutting bee females died when they cut leaf pieces from Temik-contaminated foliage, but only at excessive dosages and only during the first 10 days after treatment.

◆

Some persistent insecticides have a leaf-piece contamination hazard to alfalfa leafcutting bee females.

10.3 - Leaf pieces in a leafcutting bee nest can kill bees.

11 OTHER CONTAMINATION EFFECTS

Synergism

Combinations of Chemicals
May Be More Toxic to Bees

Synergism occurs when the toxicity of two or more compounds is greater than that expected from the sum of their effects when applied separately. Synergism is greater than additive. As an example, 3 plus 3 = 10. We've tested many two-material and three-material combinations of insecticides and miticid for their toxicity to honey bees and wild bees. **In many cases, synergistic hazards increase beyond what we expected from the individual materials.**

When Comite (propargite), Kelthane (dicofol) or Tedion (tetradifon) are added to one or more insecticides in spray mixtures, they greatly increase the toxic hazard to bees. Such action appears to be true synergism, since these specific miticides are essentially non-toxic when used alone.

◆

Synergism increases bee toxicity of pesticide mixtures.

Adjuvants, Stickers, and Spreaders

Spray Additives Make Many
Insecticides Less Hazardous to Bees

Growers and applicators use a number of different materials to increase or enhance the activity of insecticides against insect pests. Sometimes the chemicals are formulated with the insecticide and other times they are sold separately and then dumped into the spray tank to make insecticides spread better, adhere to plant surfaces tighter, or extend the residual activity. They may make some insecticides safer to bees.

Addition of **solvents and oily substances** to spray materials tends to make them safer to bees. For example, the 2 pounds per gallon Systox (demeton) formulation, which contains about seven times as much xylene as the 6 pounds per gallon formulation, is safer for this reason. In South Africa, the addition of 0.5% mineral oil to parathion sprays on blooming citrus reduced mortality among honey bees by approximately 50%. Pennwalt Superior Spray Oil added to Sevin (carbaryl) decreased mortality of bees. Also, the residual hazard of a suspension of Sevin (carbaryl) in oil applied at ultra-low volume was much shorter than that of carbaryl wettable powder.

Apparently, these oily substances increase sorption of insecticides to plant surface tissues and decrease the hazard to bees. Or, it may be adding oil hinders the penetration of insecticides from the lumen into the epithelium of the honey sac, from where diffusion into the blood occurs. Accumulation of lethal amounts of the poison in the nervous system may thus be prevented.

Acidifiers which increase the effectiveness of Dylox (trichlorfon) against pest insects do not increase the hazard to bees, except at excessive rates. Acidifiers used at more than 1:400 dilution increase the hazard of materials such as Dylox to alfalfa leafcutting, alkali, and honey bees. Also, highly acidified Spur (fluvalinate) increases alkali bee mortality. At pH 6 there is no increased hazard to bees.

Foam additives reduce problems of spray drift. We tested foam spray additives and foam spray nozzles and found no increase in hazard of several insecticides to either leafcutting bees or honey bees.

Surfactants, added to spray tanks, decrease the surface tension of water and make the insecticide mixture spread evenly over plant surfaces. Low dosages of surfactants added to sprays aren't toxic to bees. However, **nonionic surfactants,** in low concentration such as 25 ppm, in small ponds and puddles, cause extensive drowning of bees collecting water. Also, several surfactants repel bees when added to pond water at concentrations of 500 ppm.

A new formulation of carbaryl, Sevin XLR, is much lower in hazard to bees than all previous flowable formulations of this chemical. Use of much smaller particle size and addition of a latex sticker contribute about equally to the safening process.

Recent tests with the **proprietary stickers** Bond, Sur-Stik and BioFilm in combination with different insecticides gave promising results in lowered toxicity to bees. Adding a sticker reduced the bee mortality for 6 of 15 highly toxic, for 4 of 6 moderately toxic, and for one of the 3 low toxicity materials evaluated. An insecticide causing more than 25% mortality in 8 hour residue bioassay tests is usually too toxic to bees to be used on blooming crops. Adding a sticker to naled, acephate, fenvalerate, and formetanate reduced the 8 hour mortalities below 25%.

◆

Additives reduce bee hazard of some insecticides.

Sublethal Effects

Insecticides May Affect Bees Without Killing Them

Acute bee mortality (LD_{50}) studies don't show what may happen to bees from sublethal doses of an insecticide. With some insecticides, bees may be affected but not die. These sublethal effects are difficult to observe and document and data on their occurrence is scarce.

Sublethal doses of parathion alter foraging activity by slowing flight speed and affecting the time sense of foraging bees. Communication of direction and distance information by dancing bees is also adversely affected. Sevin (carbaryl) and Cygon (dimethoate) exposure may result in malformed adults. Both malathion and diazinon can shorten the life of worker bees. Malathion has less effect than diazinon. Colonies fed low doses of methoxychlor reared less brood and consumed less pollen and sucrose syrup. Also, colonies fed methoxychlor or malathion were particularly susceptible to invasion by wax moth.

Sublethal doses of Ambush or Pounce (permethrin) caused abnormal behavior of honey bees. In a classical conditioning experiment all honey bees surviving exposure to 6 pyrethroid insecticides (fluvalinate, fenvalerate, permethrin, cypermethrin, cyfluthrin, flucythrinate) had reduced learning or memory loss. Apparently, these toxicants attack the nervous system of bees and impair their response to scent stimulus. Evidence suggests that the inhibition of foraging bees resulting from an application of Ambush or Pounce is due to the sublethal effects and impaired memory. Other insecticides may also cause sublethal effects.

◆

You may not be able to see sublethal poisoning damage to bees.

Chemical Immunizers

Little Known About Chemical Immunizers & Bees

Testing of chemical immunizers to reduce bee poisoning hasn't received much research attention. In the USA, atropine sulfate and several materials tested as antidotes for bees against Azodrin (monocrotophos), parathion and Sevin (carbaryl) poisoning didn't reduce the mortality rate, regardless of method of application or whether administered before or after insecticide treatment. Several enzyme-inducing materials, including chlorcyclizine, tested for protection of alkali bees from parathion and alfalfa leafcutting bees from Sevin (carbaryl) didn't decrease the susceptibility of the bees to the insecticides. Also, honey bees fed on treated syrup containing up to 10,000 ppm P2S were not protected from Sevin sprays applied later. In Libya, honey bees fed with pyridine-2-aldoxine (2-PAM) two hours before being fed parathion or paraoxon resulted in reduced bee mortality.

◆

No antidote is available for reducing bee kills.

Repellents

The Search for an Ideal Repellent Goes on.

Repellents for reducing hazard of insecticides to honey bees have been investigated for more than 90 years. The search is for a chemical to dump in the spray tank with a highly toxic insecticide, then spray the concoction onto bloom without harming bees because they don't visit the treated field.

An effective bee repellent must be strong enough to overcome the natural plant attractiveness and prevent honey bees from foraging on plants treated with a toxic insecticide. Also, it must not harm any part of the plant. The ultimate repellent would keep enough bees from foraging the crop for enough time to significantly reduce the hazard to the colony. For many insecticides, 24 hours of reduced bee visitation might be enough. Carbolic acid was recommended by some as a bee repellent during the early 1900's; however, in field studies carbolic acid repelled very few bees. In feeding studies during the 1920's, cresol compounds and carbolic acid showed a great deal of repellency. Also, carbon disulfide, nicotine sulfate, lime sulfur, and napthalene showed bee repellency. These materials have limited use because they repel bees for only a few hours which is not enough time to reduce bee hazard significantly.

During the 1970's, laboratory screening tests of 143 chemicals in California showed compounds containing nitrogen, short sidechain-substituted phenyl acetates, and tolyl derivatives had great promise. Aromatic five-six-, and seven-membered ring structures containing nitrogen in the ring, straight-chain amides, and phenyl ring structures with short-chain-length amide substitutions were among the most repellent compounds tested. Many of these need to be thoroughly field tested.

In Israel, 20 repellents tested proved ineffective because high temperatures and intense sunlight rapidly evaporated and caused the chemicals to break down, rendering them ineffective. Also, materials such as carbolic acid and creosote added to insecticidal sprays didn't completely repel bees from the treated fields. These materials proved ineffective for reducing bee poisoning. In Egypt, the same materials reduced bee activity on cotton for only a few hours.

Molasses mixed with malathion in fly sprays was repellent to honey bees and alkali bees for a limited time.

Results with R-784 (hydroxethyl octyl sulphide) in experimental plots up to an acre were quite promising under a variety of conditions. However, large-scale tests in California on alfalfa and in New York on buckwheat showed that R-784 was not effective enough to safeguard bees against poisoning when most of the available bee forage plants in a sizable area are treated. In Chile it was used to protect *Megachile* bees from poisoning in alfalfa seed plots with some effect.

Some insecticides repel bees or more specifically applications cause reduced bee visitation to the crop.

Many years ago we established that the insecticide Systox (demeton) is an effective honey bee repellent. Even at one quarter of the normal rate used for insect pest control it repels honey bees from bloom. Also, DDT repels bees off of white Dutch clover for 1-2 days, but not other crops. In New Zealand DDT reduced honey bee visitation to crops without harming bees. But, bumble bees were not repelled by DDT. Lannate (methomyl) reduces bee visitation on red raspberry much more than it does on corn. This is probably explained by nectar contamination in the case of red raspberry. Lorsban (chlorpyrifos) reduced honey bee visits to red raspberry by 50% for several days. It seems as if most insecticides including inorganics, chlorinated hydrocarbons, organophosphates, and pyrethroids have reduced bee visitation at one time or another.

For a variety of reasons, no insecticide has been exploited as a bee repellent to reduce bee poisoning. A major reason is that they are insecticides. Applicators applying one insecticide don't want the added cost of another chemical. Another major obstacle involves registering an insecticide as a repellent on a crop. In the USA this hurdle would have to be taken up with the Environmental Protection Agency. Also, synergism could make the mixture more toxic to bees and increase bee hazard. Recently, we found the synthetic pyrethroids, permethrin (Ambush, Pounce) and fenvalerate (Pydrin) reduced bee visitation to treated crops without harming bees. Tests of permethrin on pollen shedding corn, blooming red raspberry, blooming alfalfa and

◆

blooming dandelions showed significant reduced bee visitation. The actual mechanism causing this reduction is not clearly understood, It may be true repellency by causing a new odor keeping the bees from visiting the crop. Or, more likely, sub-lethal poisoning affects the behavior of scout bees and no recruits are attracted to the crop. Permethrin (Ambush, Pounce) sprayed at night or early morning doesn't kill bees. Although spraying done when bees are flying, kills bees before the repellent effect takes place.

In California, applications of permethrin, (Ambush, Pounce) to cotton and corn repels bees but no repellency has been observed on corn in Wisconsin or Minnesota. Also, in laboratory tests bees are not repelled by the material.

Many if not all, insecticides affect foraging bees. We have determined 4 kinds of effects:

- minor irritation without toxic action (Dimilin);
- reduced visitation with obvious bee kill (Lorsban);
- conditioned response with obvious bee kill (Lannate), and
- repellency without measurable bee kill (Ambush, Pounce).

Resistance

No Species, Race or Line of Bees is Resistant to all Insecticides

It would be nice to have honey bees that were resistant to all insecticides. Then bee poisoning would cease to be a problem. But, unfortunately, no such bees exist.

In insects, including bees, four types of heritable resistance to insecticides occur:

- metabolic degradation of the poison by enzymes
- reduced poison penetration through the cuticle
- sequestering of the poison in special glands
- insensitivity or knockdown resistance

We know very little about resistance mechanisms in bees.

In metabolic degradation various enzyme systems within the insect detoxify the poison before it harms the bee. Body hairs and differences in the cuticle (outer skin) exist and can cause changes in penetration and toxicity. Also, relative resistance among insects often occurs because the insect, biochemically, isolates the insecticide within itself without harm. Genetic stocks of honey bees with high levels of tolerance to certain insecticides have been developed under experimental conditions. These lines have not been made available to beekeepers or growers. Unfortunately, resistance to one insecticide doesn't necessarily protect the bees from other insecticides. And, the resistant line would be lost if the bees were killed by a different insecticide. Incorporating many desirable genetic characteristics, such as gentleness, honey production, non-swarming, and pesticide resistance, into a stock of honey bees is difficult and expensive. In a study of the tolerance of 18 honey bee colonies to 4 different insecticides it was concluded that selection based on existing polygenic variation could provide only minor increases in resistance, and that useful developments require major gene mutations.

Queens and drones showed 4 times the resistance to DDT compared to worker bees.

Some undirected selection may take place in the field due to repeated insecticidal treatments. In California, colonies gradually changed in their susceptibility to a given dose of DDT from 65% to 15% kill over an 8 year period. Also, an increase in resistance was found to chlordane, toxaphene, and methoxychlor, but not to aldrin, dieldrin, heptachlor, lindane, organophosphates or carbamates. One commercial queen breeder developed about a 50% resistance to Sevin in many of his colonies in California.

Preliminary studies with alfalfa leafcutting bees indicate that there are differences among populations in susceptibility to Dylox and Spur.

◆

Honey bees, unlike many pest insects, haven't developed resistance to insecticides.

12   THE SCIENCE OF BEE POISONING

Combination of Lab and Field Tests Evaluates Bee Hazard

The ideal pesticide controls a specific pest without harming bees. It seldom exists so we look for the best compromise by investigating the **relative toxicity** of various pesticides to bees.

relative toxicity

Laboratory and field work must be conducted separately to evaluate materials. Difficulties sometimes exist in interpreting results of field trials when fast-acting compounds are tested; for example, poisoned bees may not return to the hives. Also, direct contact toxicity tests are often of little value as guides to residual hazard.

In California, Larry Atkins developed a useful rule of thumb method of determining the expected kill from different dosages of an insecticide. Some exceptional pesticides don't conform closely to this rule of thumb. In the laboratory the LD_{50} — the amount of pesticide which kills 50 percent of the bees tested — is determined. The LD_{50} in micrograms per bee can usually be directly converted to the equivalent dosage in pounds per acre. Known LD_{50} values can be substituted in a prepared

LD_{50}

table to obtain the anticipated percent mortalities for an appropriate range of dosages. For example, the LD_{50} of parathion is 0.175 µg/honey bee, we would expect that 0.175 lb/acre of parathion would kill 50 percent of the bees foraging in a treated crop at the time of treatment. This illustrates the basic principle that the property of a pesticide being toxic or nontoxic is determined by its dosage. Any pesticide normally considered toxic always has a dosage below which it causes no harmful effect.

We've developed standardized procedures for assessing the residual hazard of chemicals by applying pesticides to small plots of alfalfa. Bees are exposed in the laboratory to field-weathered residues on foliage samples taken from treated plots. This is followed up by small (1/2 to 1 acre) field work and then large scale tests. For certain types of chemicals bees are fed the pesticide either in the laboratory or to the colonies. Also, we collect bees from the treated bloom with a vacuum device and hold them for mortality determination.

◆
Appendix III outlines a sequential system for evaluating bee hazard of pesticides.

Analyzing Poisoned Bees

Insecticide Residues in Dead Bees Proves Pesticide Poisoning.

Many times the most common symptom of bee poisoning is piles of dead bees in front of hives. Or,

12.1 - Anywhere from a few to thousands of dead and dying bees at the entrance may indicate a pesticide loss.

12.2 - A sample of 200 dying or recently killed bees should be frozen immediately after collection. This sample should be held for instructions from the agency or firm doing the residue analysis.

absence of any field force. This is not definite proof of poisoning since not all dead bees or colonies have been killed by pesticides. Definite proof is established by chemical analyses of dead bees. Suitable chemical analyses are often difficult and costly.

Steps for Collecting Bees

1. Dead bee samples for analysis must be fresh and large enough (200 bees) to process. Dead bees more than several days old or partially decomposed are useless for analysis.

2. Collect the dead and dying bees in bags or other containers and freeze until shipped to a laboratory for analyses. Do not ship dead bees in envelopes or unprotected against being squashed in the mail. Use a cardboard container.

3. In addition, exhaustive surveys must often be conducted in the locality of the bee kill to determine which pesticide may have killed the bees. The chemist must be presented with one prime suspect to search for in the dead bee sample.

4. Dead bees dry out quickly and lose 50-80% of their fresh weight in 2-4 days. It is a good idea to count the number of dead bees in a sample. Then, results can be compared to LD_{50}'s determined in the laboratory in a more meaningful way. The fresh weight of worker honey bees averages 128 mg. Chemical residue analyses are in µg/g(ppm). By dividing by 7.8, residues can be compared to the LD_{50} given in µg/bee.

For example	the ppm residue =	the LD_{50} is
parathion	1.37	0.175
malathion	5.67	0.73
carbaryl	12.01	1.54

In one test, a fresh bee sample obtained the morning after an experimental field application of dieldrin compared with a sample of dead bees treated with dieldrin in the laboratory at LD_{95} showed twice as much in the laboratory sample ($1.0\,\mu g$/bee vs $0.5\,\mu g$/bee). When shipping bees for chemical analysis, supply as much information as possible. The following details help:

- name and address of sender
- locations of hives
- what kinds of crops or plants are near the hives
- number of colonies affected, out of total number
- starting date of symptoms
- condition of colonies affected, out of total number
- condition of colonies at last examination
- symptoms (piles of dead adults, dead brood, queenless, etc.)
- pesticide applications

Residues of pesticides may be present as the original pesticide or as an identifiable degradation product, or both. Pigments and waxes in dead bee material present terrific problems to the chemists. Such materials interfere with the analytical procedures and must be removed. The clean-up processes often prevent any positive findings which might otherwise have been obtained.

Chemists use various methods to diagnose the presence of pesticides in bees. Chromatographic methods include thin-layer chromatography (TLC), gas chromatography (GC) and gas liquid chromatography (GLC). Measurement of residual AChE activity is an indicator of poisoning by organophosphates and some carbamates. This method may give inconsistent results due to variations in the normal enzyme level in bees, and because there may be partial recovery of enzyme activity after death. Nor does this method give information identifying the actual poison involved. Kits are now available to do this by the beekeeper. Kits may be obtained from Enzytec, 8805 Long, Lenexa, KS 66215. They are useful as a starting point in investigating a bee kill.

Laboratories Possibly Able to do Pesticide Analyses of bee samples.

Many state departments of agriculture are equipped to conduct pesticide analyses for pesticides in bees and are usually willing to do so. In other countries, contact the federal government's agriculture or natural resources agency. When a bee kill occurs, collect the bees as described earlier and contact the department of agriculture for residue analysis. There are private laboratories that can do pesticide analyses and a partial listing follows. *Remember, contact these labs before you ship samples to them.* *Determine if they still conduct the service you need, and establish- in advance - what the lab costs will be. The lab will also give you shipping instructions.*

Alaska
Northern Testing Laboratories
600 University Plaza West, Suite A
Fairbanks, AK 99701
(907) 479-3115

Alabama
Southern Research Institute
P O Box 55305
Birmingham, AL 35255
(205) 323-6592

Mid-South Testing
P O Box 1303
Decatur, AL 35602
(205) 350-0864

Idaho
Northern Engineering & Testing
P O Box 7867
Boise, ID 83707
(208) 377-2100

Illinois
Aqualab
850 Bartlett Road
Bartlett, IL 60103
(312) 289-3100

Bellview Labortories
1115 Sill St.
Belleville, IL 62221
(618) 235-3600

Ohio
Microbiological Laboratories
9593 Page Road
Streetsboro, OH 44240
(216) 626-2264

Agrico Chemical
Box 639
Washington Courthouse, OH 43160
(614) 335-1582

Oklahoma
Metlab Testing Services
6625 E. 38th Street Fairbanks AK 99701
Tulsa, OK 74145
(918) 664-7767

National Analytical Labs
P O Box 9857
Tulsa, OK 74157
(918) 446-1162

Oregon
Century Testing Labs
P O Box 1174
Bend, OR 97701
(503) 382-6432

Columbia Laboratories
P O Box 40
Conbett, OR 97019
(503) 375-3387

Pennsylvania
ASW Environmental Consultants
1701 Union Blvd.
Allentown, PA 18103
(215) 434-1870

Washington
Am Test
4900 9th Avenue NW
Seattle, WA 98104
(206) 783-4700

Northwest Environmental Services
911 Western Ave., Room 336
Seattle, WA 98104
(206) 622-8353

13

LAWS AND REGULATIONS

The Label is the Law

13.1 - Pollen-shedding corn poses a high hazard to bees; laws have been passed in many areas to prevent bee kills.

When major bee kills occurred after World War II beekeepers and growers fell into two sharply divided camps. Growers and pesticide producers felt the death of bees was a small price to pay for modern agriculture. On the other hand, beekeepers, often in a state of shock from piles of dead bees and empty colonies, wanted to ban pesticides. After some 40 years, both sides now see that the other's viewpoint contains some validity. Out of the conflict came legislation and regulations to protect bees and account for the rights of the grower to protect his crops. England, Sweden, New Zealand, India, East Africa, Canada and the USA, among others, have developed regulations to protect bees.

Basically, legislation tries to keep highly bee-hazardous materials off of blooming crops or weeds. In some cases outright bans on a particular pesticide for a particular situation proved helpful. For example, Sevin (carbaryl) was not allowed on pollen shedding corn in certain areas of Washington. No case can be made to establish an outright ban on a highly hazardous insecticide. But local regional bans, conditioned by timing and bloom, help reduce bee kills. Penncap-M, a useful insecticide in orchards after bloom, shouldn't be allowed in some areas because of blooming dandelions and other plants on the orchard floor. Legislation provided for toxicity ratings of insecticides, accounted for drift and made users aware of bee kill problems. Also, legislation required pesticide labels to have bee warning statements if hazardous to bees.

In the USA, all pesticides have to be registered with the federal government before they can be sold. Before registration, chemical companies present data showing what pests the pesticide controls, the intended use, and whether it injures plants or animals when applied as directed on the label. The government also requires data on the bee toxicity and hazard of the material. All these regulations tend to afford some protection to bees and to reduce the hazards of pesticide contamination of marketable food products.

Most people who apply pesticides must pass an examination or otherwise demonstrate their knowledge on proper use of pesticides. Most states give educational courses designed to provide information for passing the test. Bee safety is taught as part of the program. Certification under this act has benefited the beekeeping industry through education and more judicious use of pesticides.

13.2— Most states provide routine instruction and examinations for pesticide applicators

In addition to federal laws, many states have laws to help protect honey bees from insecticides. The laws regulate what chemicals can be applied and under what conditions to protect bees. Some states have rather complex, involved regulations.

In California, regulations require applicators using pesticides highly toxic or moderately toxic to bees to obtain approval of the Agricultural Commissioner prior to starting spraying. Also, regulations require that all owners of apiaries located within one mile of the area to be treated be notified, at the beekeepers expense, of the intent to use a pesticide known to be harmful to bees; provided the beekeeper has made a request in writing to the Agricultural Commissioner for such notification. The beekeeper must be allowed up to 48 hours to move or otherwise protect his bees. Some things laws regulate:

- amounts of insecticide
- amount of bloom present
- temperature
- time of application
- wind velocity
- notification of beekeepers
- kind of insecticide
- registration of location of bee yards

Regulations regarding beekeepers also help to reduce bee poisonings. Many states require beekeepers to register their colonies and provide marking so others know who owns the colonies. Often apiaries must be identified by placing a conspicuous sign stating the owner's name, address, and phone number.

In most areas of the world, harm to honey bees is prevented by regulating the use of pesticides dangerous to bees on flowering crops. Usually the hazard of pesticides to bees falls into:

- highly toxic and dangerous
- moderately toxic and moderately dangerous
- non-toxic and non-dangerous

An individual pesticide may fall into 2 or all 3 groups depending on dosage and environmental conditions.

Litigation

Legislation provides that bee kills may be classified as legal or non-legal. If bees are killed by pest control measures that don't abide by the law in force, questions of liability arise. The court enters the picture to determine if there was an illegal application and the amount of compensation.

Generally, users of pesticides are held liable for any damages caused to the property or interests of others from the misuse or careless application of pesticides.

In the USA, federal legislation controls the labeling and use recommendations for pesticides. The Environmental Protection Agency uses a standard testing protocol to determine which pesticides should be approved for use on blooming plants and the conditions of that use. For example, a highly hazardous material will have a bee warning statement saying, "Do not apply nor allow this material to drift on blooming plants where it may kill bees."

Sometimes a beekeeper's only recourse is through the legal system, though many are reluctant to encourage lawsuits. Lawyers get a good share of the settlement. Lawsuits may create feuds between beekeepers and between beekeepers and farmers. For example, in 1986, in one close-knit farming community in Washington, a beekeeper sued the local aerial applicator for a bee kill caused by a mis-application of an insecticide. The lawsuit created extreme animosity between the beekeeper and his fellow beekeepers and between the farming community and all beekeepers. Ban the bees became a very popular slogan. But, if you feel you must, here are some pointers.

1. Perform frequent inspections for bee kills in apiaries located near high probability, high toxicity pesticide application crops.

2. If a bee kill is discovered, notify a state apiary inspector and the state department of agriculture, requesting a bee damage inspection and investigation.

3. Immediately start taking photos of anything associated with the bee kill.

4. Obtain dead bee samples. Have the samples chemically analyzed in order to determine insecticide residues in the bees. Timing is critical since residues dissipate after a few days. Obtain duplicate samples.

5. If the bee kill is substantial, consider retaining an expert consultant for investigative and chemical analysis purposes.

6. Contact any persons in vicinity of bee kill to get data on commercial insecticide applications in the area.

7. In some states, the state department of agriculture investigates alleged bee kills and commercial applicators potentially involved. Be prepared to help them if asked.

8. When chemical analysis results test positive for a specific insecticide residue, obtain copies of all advertisements, promotional materials, and other data pertaining to the chemical. Obtain and read the label. Also, information can be obtained from the state university

◆

Consider human interactions as well as dollar settlement when considering litigation.

♦

Only the whole bee industry working together can help prevent bee losses to insecticides.

Cooperative Extension. Standard of care by applicator comes into play.

9.   Move rapidly as possible, preserving the evidence — bee samples, colony samples, and reports.

10.   Investigating bee poisoning cases requires timeliness and precision; it would be advisable to contact a lawyer with expertise in these types of cases in order to be better prepared to attempt settlement, and if settlement discussions fail, it may be necessary to file a lawsuit.

Indemnity

During the early 1970's the US federal government sponsored a program wherreby beekeepers were reimbursed for bee losses from pesticides. The money for the program came from the federal government. The program achieved some of its objectives, mainly curtailing the disintegration of the beekeeping industry. Many beekeepers who would have been driven out of business by poisoning losses were able to stay in business and replace worn-out equipment and increase both quantity and quality of their honey bee stocks. Unfortunately, the program was terminated in 1977 by the federal government. Some beekeepers abused the program, one factor leading to termination.

MISCELLANEOUS POISONING PROBLEMS

14

A Variety of Limited-Scale Pesticide Problems Have Developed

Unique pesticide-bee conflicts have developed in many countries. Here are some interesting cases:

Don't use DiSyston G (disulfoton), methyl parathion, Phosdrin (mevinphos), TEPP and Thimet G (phorate) near alfalfa leafcutting bee shelters, alkali bee nest sites, or honey bee apiaries because of possible **fumigation hazards**, especially during warm weather.

Bees are temporarily inactivated by direct contact with oil sprays and some loss may occur.

Vapors from Vapona (dichlorvos) slow-release strips absorb into beeswax and remain a toxic hazard to honey bees for some time. Chlordane vapors also have a strong affinity for beeswax, and render combs toxic to bees. Bee losses have been associated with foragers collecting mineral feed meal containing Ronnel (fenchlorphos).

Proprietary sugar-base fly spray materials containing Dylox (trichlorfon), or Vapona (dichlorvos) and Ronnel (fenchlorphos) are hazardous to bees. We recommend mixing the commercial material with molasses (which is repellent) instead of sugar.

Mirex baits used for imported fire ant control are not hazardous to bees.

Large Scale Spray Programs

Mosquito, Grasshopper & Forest Insect

14.1— Many bee kills resulted when mosquito-control programs were conducted to halt the spread of diseases of man.

In Africa, sprays to control tsetse flies can be very hazardous to bees. Especially when highly hazardous materials such as dieldrin are used. Sometimes killing of bees occurs from insecticides applied to non-crop pests — mostly from mosquito or housefly control projects. Mosquito abatement operations involving thousands or even millions of acres have been conducted in the United States for 30 years. Baytex (fenthion), DDT, ULV sprays of Baygon (propoxur), Dibrom (naled), Lorsban (chlorpyrifos), pyrethrum, and Vapona (dichlorvos), aren't likely to cause significant kills if applied at proper rates and when bees aren't actively foraging. ULV malathion, and malathion have been relatively harmless to bees when properly applied though heavy bee losses have occurred with both when applied during the day. We tested Baygon (propoxur), Baytex (fenthion) and Lorsban (chlorpyrifos) at mosquito abatement rates and found they lost their bee hazard within a few hours.

One of the common causes of honey bee losses in mosquito abatement programs is use of incorrect dosage. If ULV malathion is used at the grasshopper control rate of 8 fluid oz per acre instead of the mosquito abatement

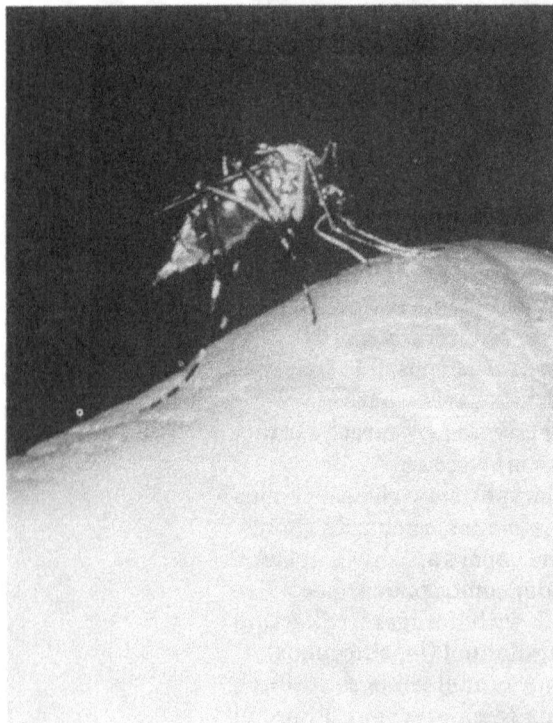

rate of 3 fluid oz, considerable bee poisoning may occur. Undiluted or ultralow volume (ULV) technical malathion spray treatments can retain a high residual toxic hazard to honey bees for at least five days and to alfalfa leafcutting bees, for at least seven days. Thirteen and one-half million acres of land along the Gulf Coast of Louisiana and Texas were treated for control of the mosquito, *Culex tarsalis* during an epidemic of Venezuelan equine encephalitis in 1971. Three fluid ounces of ULV malathion per acre were applied by airplane between sunrise and 10 AM on each treatment date. Minimal bee kills were recorded during this entire campaign. Greater problems have occurred in smaller municipal operations against St. Louis encephalitis in the Southwest. Such projects have often involved daytime applications and more hazardous materials such as Dibrom (naled). Mist sprays of Dibrom, pyrethrum and malathion combinations used to kill adult mosquitos are quite hazardous to bees. In Wyoming, Baytex (fenthion) had relatively little adverse effect on bees in a large-scale mosquito abatement program.

◆

Bee kill can occur from pesticides used in agriculture and in non-agriculture.

Grasshopper control programs in western USA often involve the treatment of millions of acres during a single season. In 1965, the federal Animal and Plant Health Inspection Service (APHIS) established ultra-low-volume (ULV) malathion sprays as the standard program. Although water-diluted malathion sprays at 1 pound ai per acre are relatively low in hazard to bees, ULV applications of 8 fluid oz (0.6 pound ai)/acre proved highly hazardous. In some areas APHIS uses a low dosage of Sevin instead of ULV malathion and they are currently investigating a microbial insecticide for grasshopper control.

Currently, bee poisoning problems are minimized by an organized, cooperative program involving APHIS, state personnel, and beekeepers. Months before grasshopper treatments are to be applied, beekeepers are alerted as to specific areas where APHIS expects to conduct campaigns during the coming season. As grasshopper populations develop and control details are finalized, beekeepers operating in localities to be sprayed are again notified concerning details of time and place. In this way, 1.08 million acres of rangeland were treated in Washington in 1973 and another 500,000 acres in 1979 without significant bee losses.

Prior to 1985, APHIS conducted programs only on rangeland and no closer than 1 mile to crop land. This helped reduce potential bee poisoning problems. In 1985, APHIS began treating land bordering crops and serious bee poisoning problems occurred in Idaho in 1985.

Large-scale insect forest defoliator control programs have been conducted periodically in the United States since 1947. Early programs involved DDT without hazard to bees. Sevin (carbaryl) first caused widespread losses of honey bees in New York gypsy moth control programs in 1959. More recently, Sumithion (fenitrothion) has been used against spruce budworm in the East and Sevin against hemlock looper in the West. DDT was used one last time for Douglas fir tussock moth control on 460,000 acres in the Pacific Northwest in 1974. The most promising material for future defoliator control is Dimilin (diflubenzuron). Low-hazard, viral and bacterial, microbial insecticides are being tested for tussock moth control and these are non-hazardous to bees.

◆

Stay aware of potential large-scale insecticide spraying operations in areas where you keep bees.

Cooperation Reduces Bee Poisoning

Beekeeper - Grower Cooperation is a Key

Better cooperation between beekeepers and growers helps reduce bee poisoning problems. In many cases, a grower, simply through ignorance of the bee hazard of an insecticide, has caused tremendous damage to a large number of colonies. The timing or insecticide could have been modified so that little or no poisoning occurred without unduly inconveniencing the grower.

Beekeepers need to get acquainted with the farmer on whose land they place bees. Learn about his pest control practices. Be aware of other special problems he might have which might harm bees. Be able to suggest to growers potential bee poisoning problems.

Never set bees down without permission of the land owner. Even in Washington in 1987, beekeepers unloaded bees on land without asking permission. Also, a few beekeepers ask permission, set off the bees and leave never to return until time to gather honey. Take an interest in the farmer and he will take more of an interest in your bees. With colonies hired for pollination, a written agreement executed between the parties is nice and business like. Such contracts should include details of the responsibility of the beekeeper in providing strong and effective colonies and of the farmer in safe-guarding the bees from insecticides.

Be able to suggest to growers potential bee poisoning problems. And, even more importantly, where to go to get information to prevent bee kills.

> In modern agriculture, where the bee-keeper often depends on farmers for bee pasture, cooperation is essential.

Voluntary Early Warnings

A Non-Mandatory System Notifies Beekeepers When Sprays Will Occur

Under these systems beekeepers receive advance warning of spraying that may kill their bees. The system works only when contact can be made with beekeepers whose colonies are at risk, and when beekeepers work closely enough with the growers and sprayers to keep the scheme in operation. One such system in Glen, California involves 50 cooperating beekeepers with about 7,000 hives. Each pays $100 to operate the system. Each marks his apiaries on a large map given to the voluntary organizer of the scheme. The aerial spray operators give advance notice of their spraying to the organizer, so that she can give 48 hours notice to the beekeepers concerned. They can then move or close up their hives or decide that the risk isn't great enough to warrant any action. The group maintains an insurance policy that pays a nominal payment for the colonies destroyed or severely damaged. Arizona has a similar system.

In the Kittitas Valley of Washington a "gentlemen" agreement between beekeepers, spray applicators, and haygrowers operates through the university Cooperative Extension office. The agreement follows:

• Beekeepers will have prospective hive locations marked on a county map with their assigned apiary numbers or other distinguishing marks and distribute copies of this map to Kittitas County commercial applicators and Cooperative Extension no later than May 15 of each year.

- Hive yards shall be marked in accordance with Washington State Apiary Law or with an identification number that correlates with the map.
- Beekeepers will get landowner permission to place bees on land in Kittitas County.
- If warned of an imminent application, beekeepers will attempt to take precautions to protect their bees.
- Applicators will attempt to get fields sprayed earlier in the season. They will attempt to convince growers to accept preventative programs.
- Applicators will try to use the safest chemical that will do a comparable job of controlling pests. Economic cost of chemicals may be considered in the decision.
- When crops or weeds in crops are approaching their bloom period, applicators will try to limit applications to morning or evening to avoid periods of greatest bee activity. Fields nearing bloom will receive priority for treatment so they may be sprayed prior to reaching their bloom period.
- Growers will attempt to monitor insect populations and contact their consultants or applicators if a problem is seen.
- Special Local Need registrations on alternative chemicals will be pursued by Cooperative Extension.

◆ **In some areas early warning systems have reduced bee kills from pesticides.**

In Crete, the Government Agricultural Services run a scheme, under which they warn all beekeepers known to them, before aerial spraying of olive groves. In addition, newspapers carry a notice one week in advance of the spray. Most beekeepers move their hives out of the area. There is no compensation for any spray damage to the bees.

In England, some counties have schemes for warning beekeepers of the major spray hazard, aerial applications of insecticides to *Brassica spp.* crops. Early warning systems provide some relief to beekeepers but have several major drawbacks:

- It takes time to move or cover bees
- It is labor intensive and expensive
- There may be no safe place to move colonies to.

Also, the sheer magnitude of spraying can overwhelm beekeepers. In diversified agricultural areas at least one hazardous spray will be applied within the flight range of an apiary more than once a week.

CONTROL SPRAYING TO PREVENT BEE LOSSES.

Activity	SPRAY	NO SPRAY	SPRAY
Plant	bud	bloom	petal fall
Attractiveness To Bees	NOT ATTRACTIVE	ATTRACTIVE	NOT ATTRACTIVE

14.2 - Simple Extension Service artwork will help growers and beekeepers understand how to protect pollinators.

15   REDUCING POLLINATOR DAMAGE AND DEATH

Beekeepers Can Help Reduce Bee Poisoning

Use Low-Hazard Apiary Locations

Establish holding apiaries at least 4 miles from crops being treated with toxic materials. Select sites relatively isolated from intensive insecticide applications, and not normally subjected to drift of chemicals. Don't leave unmarked hives next to orchards or fields that might be treated; the beekeeper's name, address and phone number should be marked in print large enough to be read at some distance, so that he can be contacted when hazardous sprays are to be applied.

Use Educational Programs

We know many ways to help alleviate bee poisoning. Often, severe losses could have been avoided by relatively simple modifications of pest control programs. Teach growers, pesticide applicators, and beekeepers

how to reduce poisoning. This is the most important method of poisoning prevention.

Learn about pest control problems and programs so you can develop mutually beneficial agreements with growers concerning pollination service and prudent use of pesticides. Check to see that pollinator protection questions are included on applicator examinations.

Confine Honey Bees Inside the Hive

Honey bees can be confined inside the hive to reduce amount of bee kill. Whenever confining bees, provide plenty of water inside the hives to allow the bees

15.1 — Avoid high-risk apiary locations, such as this one near a corn field.

15.2 — Simple confinement systems work for part-time bee-keepers only.

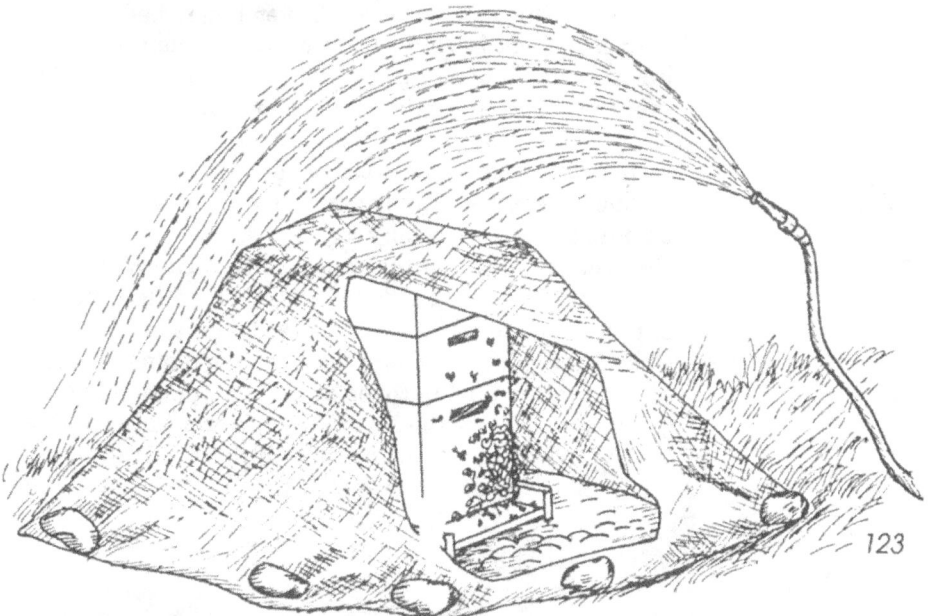

123

15.3 — Burlap covering.

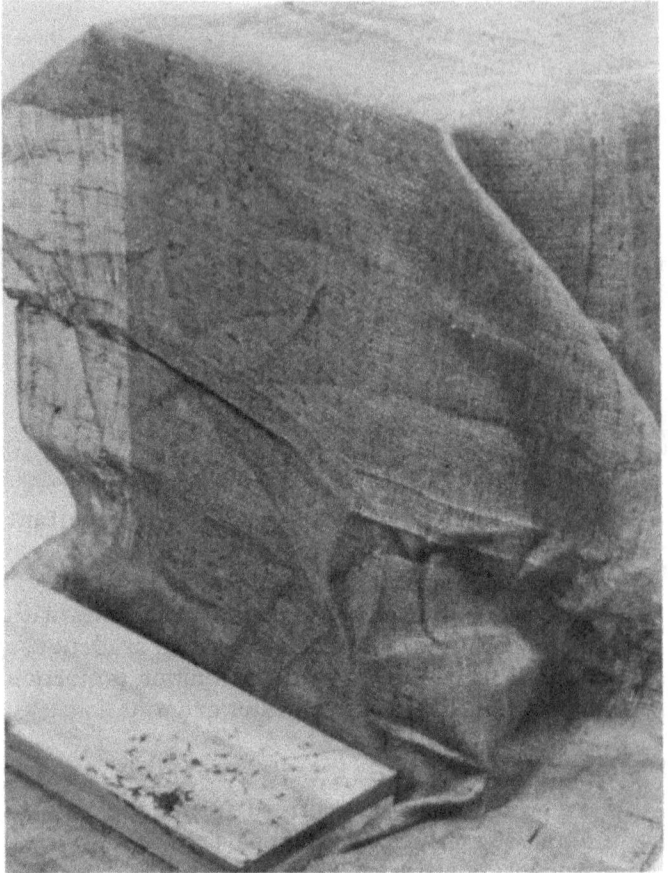

to reduce the temperature. Cover honey bee colonies with large wet burlap sacks to protect them from the initial hazards of an insecticide. Such covers should be over the hives during the night before the crop is treated. Block the hive entrance. The colonies may be covered for up to 2 days and the bees confined within the hive if the burlap is kept wet. Apply water to the burlap every 1 to 3 hours to keep the colony cool and as a water source for the bees cooling the interior of the hive. The Wardecker Waterer works well. It's basically a modified shallow super to provide a reservoir of water within the hive. Colonies equipped with this device don't need water applied to the burlap cover during confinement. Also, wet sponges inside the hive can serve as a water source.

Overheating of colonies of confined bees can lead to death of more bees than would have been killed by the insecticide, and in a hot climate such closure may well kill the entire colony. Large colonies are more liable to heat damage than small ones. The best method for honey bee protection from insecticides appears to involve covering the hives with burlap, shading, feeding pollen and syrup, using Miller bottom boards and Wardecker Waterers. But, such treatment is time consuming and uneconomical. In Wyoming, hives covered with wet sacking for one day suffered only low adult mortality from an application of methyl parathion. In Arizona, when entrances of alternate colonies were covered with wet sacks before and until 5 hours after an application of parathion, bee losses were only about 11% of those in the uncovered hives. Also, in Arizona, tests with toxaphene and methyl parathion significantly more colonies survived the treatment when they were covered with burlap and given water for 12 hours during and after spraying. In California, the number of bees flying during the day after they had been confined in hives covered with wet sacking for 24 hours was 1 to 49% higher than expected.

Screens put over the entrance work fine for confining bees in hives without leak holes. Good ventilation and a cluster chamber is necessary. And, water if confinement is more than a few hours. Ventilation must be provided at the top or sides of the hive, not only through the entrance, because it may get blocked by dead bees. Ventilating screens should have as large a mesh as possible. In Israel colonies were confined for up to 4 days without high mortality if adequate ventilation and water were provided. For confining bees inside the hive provide plenty of:

- space in the hive
- ventilation in the hive
- shade
- water inside the hive

Another way to keep honey bees inside hives during insecticide treatments is to sprinkle them with water to simulate rain. The workers think it is raining and don't forage. Water needs to be applied constantly and in enough volume to keep bees inside the hives. Usually this method isn't practical since most apiaries don't have water in a pressurized line for a sprinkler.

Confining bees is only feasible and of value with small numbers of colonies.

Pesticide-contaminated pollen carried into the hive can be a severe problem. Two actions by the beekeeper may be helpful:
- prevent pollen from being stored in the hive
- provide a safe pollen supply inside the hive.

During aerial spraying of Sevin (carbaryl), pollen trapping at the hive entrance has been successful and also unsuccessful in reducing amount of bee losses. In one test providing pollen in the hive during the tasseling period of corn greatly reduced the collection of toxic corn pollen. Provision of pollen supplement has been found useful in conjunction with other measures for colonies in cotton fields during pesticide treatments.

Moving Bees

Sometimes, prudence says to move hives outside the high-hazard area. When honey bees will be exposed to highly toxic pesticides having long residue action, move the colonies away from the treatment area. Also, if the fields are going to be continually treated it may be best to move the bees out. Repeated insecticide applications can cause severe damage.

But sometimes moving colonies short distances for 24 or more hours to avoid short-residual poison sprays may be the best solution to the bee-mortality problem.

Moving colonies, however, can be costly, not only in terms of vehicle expenses and labor, but often queens are killed or injured during the move and the morale of the

15.4 — Moving colonies is labor intensive and expensive, but may be the best answer to avoiding colony damage or death

colony may be disrupted for several days. The biggest problem with moving bees is finding pasture where the risk from pesticides is nil. Areas with low or no pesticide exposure and satisfactory honey production are difficult to find. Usually, especially in diversified agricultural areas there is no place to go. Artificial nests of *Osmia* spp. bees can be removed from the orchards during spraying operations and stored at cool temperatures for a few days. The alfalfa leafcutting bee can be safeguarded by storing the nest units in a cool room or cellar for a few days while treating the fields. Nests with females in the ends of the tunnels can be moved at night. This bee is nearly inactive at 70°F and completely inactive at 60° F. Also, leafcutting bee nest shelters can be built to be covered or closed during insecticide applications to reduce the drift of insecticides into the nest structures. When placing leafcutters on fields in a rotation plan, don't move nest shelters in until at least one week after Cygon (dimethoate), Furadan (carbofuran), Lorsban (chlorpyrifos), Supracide (methidathion), or malathion ULV treatments.

◆

Sometimes, moving bees away from dangerous situations solves the bee poisoning problem.

Treating Honey Bee Colonies

Take Steps to Help Colonies Recover from Pesticide Poisoning

If a colony loses only its foragers and has plenty of honey and pollen in the hive it will usually recover without any help from the beekeeper. If the particular insecticide has a long residual effect, weakened colonies should be moved to a safer place. In any case, if the loss occurs at the beginning of nectar flow, there will be little surplus honey and the colony will have to be fed sugar for winter stores.

If brood and nurse bees die, poisoned pollen is present in the hive. The colonies should be moved to a safer place and combs containing contaminated pollen removed from the colonies. Hive bees will continue to die as long as the poisoned pollen remains in the hive. Even when such colonies survive they may be of little or no economic benefit. Often, package bees or swarms have died when placed on combs from colonies which previously had been poisoned.

Combs with contaminated pollen can be cleaned.

Soak them in water for 24 hours, then wash the pollen from the cells and air dry the combs. Or, melt the entire comb and replace with wax foundation. Adding more bees to a colony weakened by loss of its foragers sometimes results in quicker recovery of the colony. Or else, unite weakened colonies. This often works better than trying to get all colonies back into production. To stimulate brood production feed weakened colonies sugar syrup. Also, supplemental feeding of either natural pollen, pollen substitute or combinations either in dry form, in patties, or mixed with honey syrup, sugar syrup, or divert sugar syrup, aids population recovery. Colonies can be brought back to full production after about 2 months using intensive care.

To help colonies recover from bee poisoning:

- feed pollen or pollen substitute
- feed sugar syrup
- give water
- add a package of bees (if hive is not contaminated)
- unite weak colonies
- protect from heat or cold
- move them to a pesticide free area that has natural nectar and pollen available.

Steps Growers and Applicators Can Take to Prevent Bee Poisoning

◆

Factors favoring increased brood rearing, aid colonies in their recovery.

Use Low-Hazard Materials

Insecticides toxic to bees should not be applied to crops in bloom, including adjacent crops or interplants and flowering weeds, in orchard cover crops, or field edges. With aerial application, the aircraft should not be turned, nor the materials transported back and forth across blossoming fields.

Use Low-Hazard Methods of Application

Systemic insecticides injected into the soil are safer than spraying them on plants. Fine sprays are safer than coarse sprays. Ground application is generally safer than aerial application, because of less drift of the insecticides and smaller areas are treated at one time.

Use Low-Hazard Formulations

Dusts are high hazard to bees. Wettable powders are more hazardous than emulsifiable concentrates or solutions. Adding a solvent or oil substance makes sprays safer to bees. Granular formulations are low in hazard to bees.

Use Low-Hazard Timing of Applications

Chemicals which break down within a few hours can be applied during late evening, night or early morning, with relative safety to bees. Early morning applications are more hazardous than late evening or night applications.

Remove Weed Flowers

Flowering weeds should be eliminated from orchard cover crops or field edges, by mowing or beating. Treatment with 2,4-D is the best way to remove dandelion blooms. This is especially important during a dearth of pollen and nectar plants in the area, when bees may fly for several miles in search of flowers.

15.5— Growers and beekeepers in Yuma Arizona have used night-applications with short-residual materials to significantly reduce bee losses.

Modify Spray Programs in Relation to Temperature

Insecticides should not be applied when unusually low temperatures are expected afterwards, because residues would then remain toxic to bees for a much longer time. Conversely, when high temperatures cause bees to start foraging earlier in the morning, or to continue foraging later in the evening, application times should be shifted accordingly. Also, bees may cluster outside hives and direct contact from the pesticide will likely kill them. If the temperature is expected to reach the dew point during the night, no not apply even insecticides with low hazard to bees.

If temperature is high at time of application, some insecticides may break down more quickly, making them safer earlier, but bees start foraging earlier in the morning and continue later in the day.

Use Selective Insecticides and Integrated Pest Management Programs

Encourage the development and use of selective insecticides and integrated control measures. Integrated programs which rely upon biological and cultural methods, as part of the pest management system, tend to minimize use of chemicals. Selective insecticides are often less hazardous to bees and to other beneficial insects.

Use Non-Hazardous Sites for Pesticide Dumps

Do not dump unused stocks of insecticides where they might become a bee poisoning hazard. Sometimes bees collect any type of fine dust material when pollen is not readily available Under such conditions, they may actually carry pesticide dusts back to the colony.

Use Economic Enjury Level Guidelines

Don't apply any pesticide unless the crop is so heavily infested treatment it is worth doing.

Other

When roadside and other weed control operations involve 2,4-D and similar compounds on blooming plants, select the formulations or derivatives known to be least harmful to bees. Our tests show that at maximum dosage, alkanolamine salts and isopropyl esters are more toxic than other forms. Oily formulations seem to be more hazardous to bees. Spraying in late afternoon or evening will also lessen the hazard, since bees will not visit the blooms after they become curled.

Do not apply insecticides over nesting sites of wild bees where these occur next to fields being treated. This is especially important in protecting alkali bees, bumble bees and leafcutting bees.

Contact the beekeeper and ask him to remove his colonies from the area (or keep the bees confined during the application period) when such measures are feasible. Our tests show that up to 90% of the killing of bees by parathion, for example, occurs during the first 24 hours after application. Don't move hives back into parathion-treated fields less than 1.5 days after the treatment is applied.

Learn the pollination requirements of your crop.

15.6 - Good forage locations positioned away from pesticide hazards.

131

Bee Forage Preserves

Bee Forage Preserves Provide Uncontaminated Food for Bees

Current agricultural practices of cultivation, weed control and removal of hedges cause an increasing dependence on cultivated crops for bee forage. To complete this vicious circle, bees become more and more susceptible to poisoning from crop pest-control programs. To break the cycle, the encouragement of bee forage plants in wild or waste areas isolated from cultivation seems logical.

Several years ago we sowed bee plants in the Winchester Wasteway of the Columbia Basin Irrigation Project of Washington. This area contains large amounts of land subirrigated by water drainage and seepage from the irrigated fields. It is isolated from cultivated fields and will never be farm land. Several leguminous plants made promising growth and production under minimum-care conditions (no help beyond the initial plot establishment). However, they were ultimately replaced by noxious weeds, especially Canada thistle.

Game management, plant introduction, soil conservation and natural resource interests usually give excellent cooperation, because bee forage fits into their multiple land-use and crop-improvement concepts. Another type of bee forage preserve system developed in Washington uses legislation and regulations on insecticide applications to reduce bee poisoning. They limit sweet corn production to certain periods and localities by 'zoning orders' for protection of bees. In addition, three areas of Washington have been developed as 'legume seed preserves' under this system. Bees are further protected by bee poisoning regulations on blossoming alfalfa, clover, and mint crops within the preserves.

APPENDIXES 16

Appendix I

SEQUENTIAL TESTING FOR BEE HAZARD

Toxicity and hazard are not the same.

Toxicity is the inherent property of a chemical to cause adverse biological effects. Toxicity of pesticides to bees can be defined by laboratory tests.

Hazard is the possibility of producing adverse effects in specific circumstances. The hazard of pesticide usage for bees can be assessed by field and cage tests.

Sequential testing utilizes a step system to evaluate the hazard of a pesticide to bees. In this decision making process orderly steps and techniques are used. We start with step 1 and proceed through the tests to decide the toxicity and hazard of a pesticide.

Step 1 —
Laboratory Test for Acute Contact LD$_{50}$

This step is required of all pesticides which may be used in a manner that may harm bees.

Most investigators agree that almost any spray tower, topical drop or other method of determining the inherent contact toxicity of pesticides to bees suffices. The vacuum bell-jar dusting technique developed by Anderson and Atkins is preferred by some because of the backlog of data on hundreds of pesticides already determined by this system.

Classification of toxicity based on LD$_{50}$'s (μg/bee):
- > 100   virtually non-toxic
- 11 - 100   slightly toxic
- 2 - 10.99   moderately toxic
- < 2.0   highly toxic

Europeans have developed a hazard ratio of pesticides as a method to predict hazard or classify hazard of pesticides to bees. A hazard ratio or factor is determined by dividing the recommended dose (in grams of actual ingredient) by the laboratory LD_{50}. For this scheme:

- < 50   non-toxic
- 50 to 2500   slightly to moderately toxic
- > 2500   highly toxic

Step 2 — Residue Bioassay for RT_{25} and RT_{40}

This step is also required of all pesticides produced which might be used in a manner that may harm bees.

Small (0.01 acre) plots of alfalfa are sprayed with a backpack sprayer. Foliage samples are taken from plots at 2, 8, and 24 hours after application. If mortality of bees exposed to 24 hr residues is greater than 25%, sampling at 48 hr. intervals should continue until mortality of bees exposed to treated foliage is not significantly greater than the control mortality. Bees of uniform age are confined in cages constructed from 15 cm diameter plastic petri dishes and cylindrical spacers formed from 18 x 2 inch strips of wire screen. Squares of cotton soaked with 50% sugar syrup are placed in the bottoms of the cages. Foliage samples are cut into 1-2 inch lengths and mixed thoroughly. About one pint of treated foliage is placed in each cage. Then a group of 20-60 bees, anesthetized with carbon dioxide or chilled are added to each cage.

Cages of bees are held at 75°F. and checked for mortality after 24, 48, and 72 hours. At least four cages of bees should be tested per replicate and each treatment should be replicated at least 3 times over time to reduce variability due to weather conditions.

Please see pages 68 and 159 for more information on RT's.

Step 3 — Subacute Feeding Study

Any pesticide with an acute contact LD_{50} of less than 11 micrograms per worker bee requires this test.

Investigations require package bees in new equipment. Pesticides are normally fed in sugar candy in order to reduce consumption of unequal dosages of the chemical. Test colonies are held in walk-in cages and provided water and pollen or pollen substitute. Tests run for 42 days or at least 2 complete brood cycles. Measurements include amount of eggs, open brood and sealed brood; gross colony weight; adult population and presence or absence of abnormalities or disease. Todd dead bee traps attached to colonies allow daily samplings without disrupting the colonies.

A modification involves feeding individual bees known amounts of pesticides in sugar syrup in the lab.

Step 4 — Field Testing

This step is required when data from previous steps indicate the pesticide may present a problem to bees.

Isolated field plots, use of dead bee and pollen traps, chemical analysis of bees and pollen, and use of various adult and brood measurements may be useful in such investigations. Also, behavior and number of foragers in plots is useful information. Replicated plots should be larger than one acre and for some tests, a minimum of ten acres is best.

A.1 — Bioassay for residue

A SEQUENTIAL TESTING SCHEME

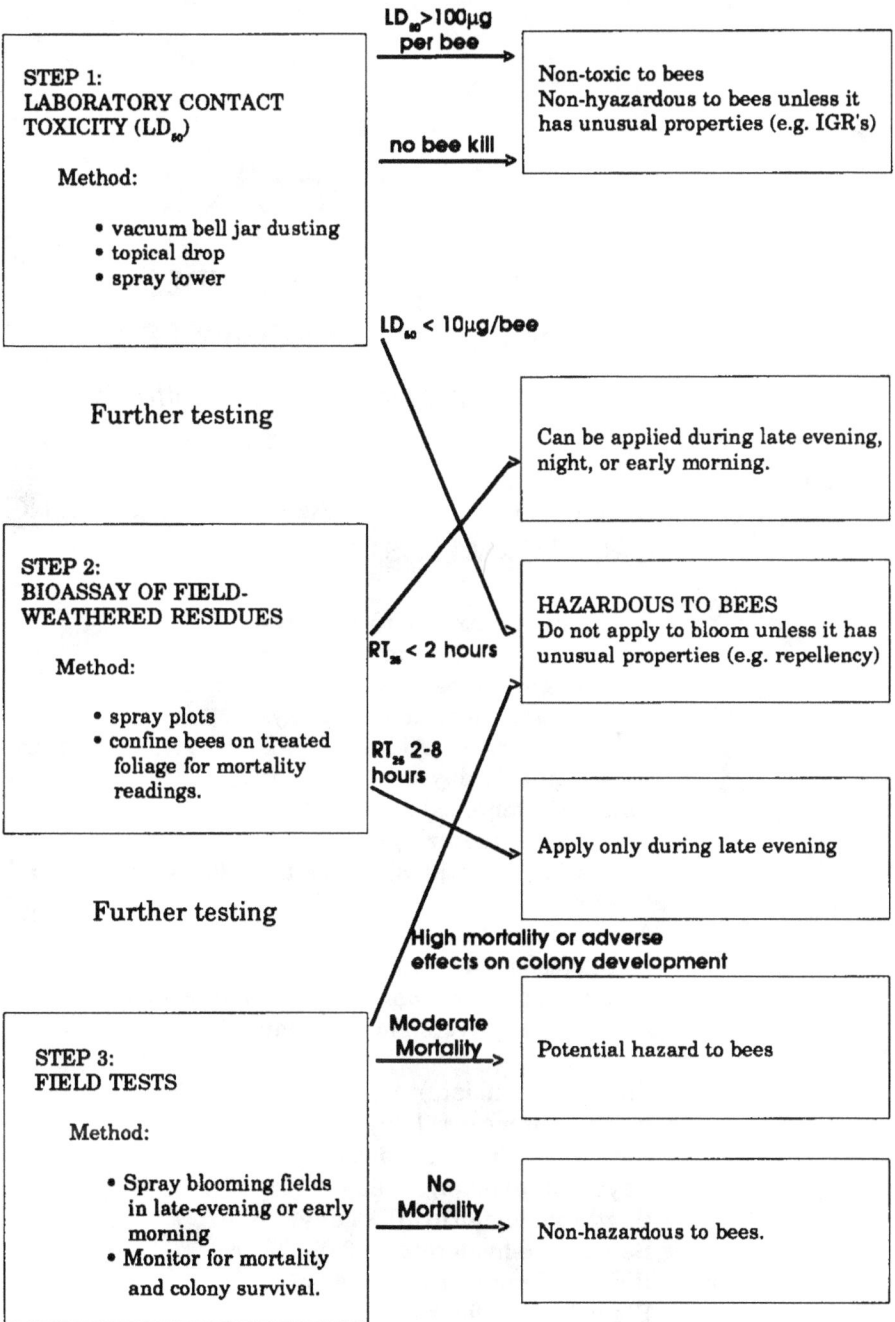

STEP 1:
LABORATORY CONTACT
TOXICITY (LD$_{50}$)

Method:

- vacuum bell jar dusting
- topical drop
- spray tower

Further testing

LD$_{50}$>100µg per bee

no bee kill

→ Non-toxic to bees
Non-hyazardous to bees unless it has unusual properties (e.g. IGR's)

LD$_{50}$ < 10µg/bee

STEP 2:
BIOASSAY OF FIELD-
WEATHERED RESIDUES

Method:

- spray plots
- confine bees on treated foliage for mortality readings.

Further testing

RT$_{25}$ < 2 hours

RT$_{25}$ 2-8 hours

Can be applied during late evening, night, or early morning.

HAZARDOUS TO BEES
Do not apply to bloom unless it has unusual properties (e.g. repellency)

Apply only during late evening

High mortality or adverse effects on colony development

STEP 3:
FIELD TESTS

Method:

- Spray blooming fields in late-evening or early morning
- Monitor for mortality and colony survival.

Moderate Mortality → Potential hazard to bees

No Mortality → Non-hazardous to bees.

Appendix II

TOXICITY OF INSECTICIDES AND MITICIDES TO HONEY BEES

Length of residual toxic effects in hours or days.

GROUP A — DO NOT APPLY ON BLOOMING CROPS OR WEEDS.

Accothion (fenitrothion) 1-5 days
Actellic (pirimiphosmethyl) > 8 hours
Advantage (carbosulfan) > 3 days
Agrothion (fenitrothion) 1-5 days
aldrin (Alderstan, Aldrex, Astex, HHDN, naldrin, Octalene) > 1 day
Amaze (isofenphos) > 1 day
Ambush (permethrin) 1-2 days *
Ammo (cypermethrin) (more than 0.025 lb/acre) > 3 days
Anthio (formothion)
Asana (esfenvalerate) 1 day *
Avermectin (more than 0.025 lb/acre) 1-3 days
Azodrin (monocrotophos) > 1 day**

Banol (carbanolate)
Baygon (propoxur) 1 day
Baytex (fenthion) 2-3 days
Baythion (phoxim) > 1 day
Baythroid (cyfluthrin) > 1 day
Belmark (fenvalerate) (>0.09 lb/acre)
Bidrin (dicrotophos) 1.5 days
Bladafum (sulfotep)
Bolstar (sulprofos) > 1 day

Bomyl 2 days
Bracklene (dicapthon)
Bromex D,WP (naled) > 1 day
Brigade (bifenthrin) > 1 day

calcium arsenate > 1 day
Capture (bifenthrin) > l day
Carbicron (dicrotophos) 1.5 days
chlorthion
Cidial (phenthoate) > l day
Ciodrin (crotoxyphos)
Colep
Curater F (carbofuran) 7-14 days
Cyflee (famphur)
Cygon (dimethoate) 3 days
Cymbush (cypermethrin) (0.02 lb/acre) > 3 days
Cynem (thionazin)

Danitol (fenopropathrin) 1 day
Dasanit (fensulfothion) 1 day
De-Fend (dimethoate) 3 days
DDVP (dichlorvos) > 1 day
Decis (deltamethrin)
diazinon (Diazitol,Basudin) 2 days
Dibrom D or WP (naled) > 1 day
Dicofen (fenitrothion) 1-5 days
dieldron (Dilstan, HEOD) 2 days
Dithiofos (sulfotep)
DNBP (dinoseb) (Basanite, DN-239,DNIBF,DNOSBP,
      DNSBP,Ivosit) l day
DNC or DNOC (dinitrocresol) (>0.4% dilution) > 1 day
Draza (methiocarb) > 3 days
DTMC (aminocarb) > 3 days
Dursban (chlorpyrifos) 4-6 days

Ekalux (quinalphos)
Ekamet (etrimphos) > 2 days
Elgetol (dinitrocresol) (1.5 qt/100 gal or more) > l day
Elsan (phenthoate) > 1 day
EPN l day
Ethyl Guthion (azinphosethyl)
Ethyl-methyl Guthion

Famophos (famphur)
Fenstan (fenitrothion) 1-5 days

fenoxycarb
Ficam (bendiocarb) > 1 day
flucythrinate
Folimat (omethoate) > l day
Folithion (fenitrothion) 1-5 days
Furadan F (carbofuran) 7-14 days

Gamma-Col (gamma-HCH)
Gammalin (gamma-HCH)
Gammexane (gamma-HCH)
Gardona (tetrachlorvinphos)(higher rates)
Garrathion D (carbophenothion) > 1 day
Gusathion (azinphos-methyl) 2.5 days
Guthion (azinphos-methyl) 2.5 days

Hamidop (methamidophos) 1 day **
HCH (gamma-HCH)
heptachlor (Velsicol) > l day
heptenophos
Hostathion (triazophos)

Imidan (phosmet) 1-4 days

Karate (cyhalothrin) > l day
Kilval (vamidothion)
Knox Out (encapsulated diazinon) > 2 days
Kotol (gamma-BHC)

Lannate D (methomyl) > 1 day
lead arsenate > l day
Lebaycid (fenthion) 2-3 days
lindane > 2 days
Lorsban (chlorpyriphos) 4-6 days

malathion D > l day
malathion ULV (8 fl oz/acre or more) 5.5 days
(Cythion,Maldison,mercaptothion)
Matacil (aminocarb) (1 lb/acre or more) > 3 days
Mesurol (methiocarb) > 3 days
methyl parathion (Metacide,metaphos,Wofatox) 1-4
    days
Methyl Trithion (methyl-carbophenothion)
Monitor (methamidophos) l day**
Murvin 'fifty' (carbaryl) > 3 days

Nemacur P (fenamiphos) > 1 day
Nemaphos (thionazin)
Nexagon (bromophos-ethyl) > 1 day
Nogos (dichlorvos) > 1 day
Nudrin D (methomyl) > 1 day
Nuvacron (monocrotophos) > 1 day **
Nuvan (dichlorvos) > 1 day

Orthene (acephate) > 3 days

Pact (thianitrile) > 1 day
Papthion (phenthoate) > 1 day
paraoxon
parathion (Folidol,Fosfex,Thiophos) 1 day
Penncap-M (methyl parathion) 5-8 days**
phosphamidon (Dicron 54 SC,Dimecron,Lirothion) 1-2
    days
Pirimicid (pirimiphos-ethyl) > 1 day
Pounce (permethrin) 1-2 days*
Prolate (phosmet) 1-4 days
Pydrin (fenvalerate) (more than 0.1 lb/acre) 1 day*
Pyramat

Rebelate (dimethoate) 3 days
resmethrin
Ripcord (cypermethrin) (>0.02 lb/acre)
Rogor (dimethoate) 3 days

Sevin WP (carbaryl) 3-7 days
Sevin-4-oil (carbaryl) (more than 0.5 lb/acre) > 3 days
Sevin XLR (carbaryl) (more than 1.5 lb/acre) > 1 day
Sinox (dinitrocresol) 1 day
Sinox General (dinoseb) > 1 day
Soprocide (gamma-BHC)
Standak (aldicarb sulfone) 1 day
Stirofos (tetrachlorvinphos) (higher rates)
Strykol (gamma-BHC)
Sumithion (fenitrothion) 1-5 days
Supersevtox (dinoseb) 1 day
Supracide (methidathion) 1-3 days
Swat (bomyl) 2 days

Tamaron (methamidophos) 1 day **
Telodrin (isobenzan)
Temik G (aldicarb) (apply at least 4 weeks before bloom)

Terracur (fensulfothion) 1 day
Tiguvon (fenthion) 2-3 days
TRI-ME (methyl-carbophenothion)
Trithion D (carbophenothion) > 1 day

Ultracide (methidathion) 1-3 days
Unden (propoxur) 1 day

Vapona (dichlorvos) > 1 day
Vigon F (dinoseb) 1 day
Volaton (phoxim) > 1 day
Warbex (famphur)

Yaltox F (carbofuran) 7-14 days

Zectran (mexacarbate) 1-2 days
Zinophos (thionazin)

* Safened by repellency under arid conditions.
** Can cause serious problem if allowed to drift into vegetable or legume seed
crops.

GROUP B — APPLY ONLY DURING LATE EVENING

Avermectin (0.025 lb/acre or less) 8 hours
Belmark (fenvalerate) (< 0.1 lb/acre) 6 hours
Bromex EC (naled) 16 hours
Dibrom EC (naled) 16 hours
Dursban ULV (chlorpyrifos) (0.05 lb/acre or less) < 2
    hours
Ekatin (thiometon)
malathion EC 2-6 hours
Phosdrin (mevinphos) < 5 hours
Pydrin (fenvalerate) (< 0.1 lb/acre) 6 hours
Savit (carbaryl) (1.5 lb/acre or less) 8 hours+
Sevin XLR (carbaryl) (1.5 lb/acre or less)
    (not > 1:19 dilution) 8 hours+
Thimet EC (phorate) 5 hours
Thiodan (endosulfan) (more than 0.5 lb/acre) 8 hours
Tiovel (endosulfan) (more than 0.5 lb/acre) 8 hours
Vydate (oxamyl) (1 lb/acre or more) 8 hours
+These materials are more hazardous to bees in a moist climate and under slow-
drying conditions

GROUP C — APPLY ONLY DURING LATE EVE-NING, NIGHT, OR EARLY MORNING.

Abar (leptophos) < 3 hours
Abate (temephos) 3 hours
Acrex (dinobuton) < 2 hours
Acricid (binapacryl)
Afugan (pyrazophos)
Ammo (0.025 lb/acre or less) < 2 hours
Aphox (pirimicarb) < 2 hours
Aramite D
Aspon (propyl thiopyrophosphate) < 2 hours
Asuntol (coumaphos)

Baygon ULV (propoxur) (0.07 lb/acre or less) < 2 hours
Baytex ULV (fenthion) (0.1 lb/acre or less) 2 hours
Biothion (temephos) < 2 hours
Birlane (chlorfenvinfos)
Bladan (TEPP) < 5 hours

Carzol (formetanate) 2 hours
chlordane (octachlor,Octa-Klor, Sydane 25) < 2 hours
Citram (Tetram)
Co-Ral (coumaphos)
Croneton (ethiofencarb) < 4 hours
Curacron (profenofos) < 6 hours
Cymbush (cypermethrin) (< 0.02 lb/acre) < 2 hours

DDT (Deestan, Didi-Col, Didimac, Vitanol) < 4 hours
DDVP MA (dichlorvos)
Delnav (dioxathion) < 2 hours
Derris D (rotenone) < 2 hours
Dessin (dinobuton)
dieldrin G (HEOD) < 2 hours
Dilan
Dimetilane (dimetilan)
Dipterex (trichlorfon) 3-6 hours
Di-Syston EC (disulfoton) 7 hours
DNOC (dinitrocresol) (< 0.4% dilution)
Dyfonate (fonofos) 3 hours
Dylox (trichlorfon) 3-6 hours

Elgetol (dinitrocresol) (1.5 pt/100 gal or less) 2 hours
endrin (nendrin) 2 hours

Eradex (thioquinox)
ethion (diethion, Nialate, Sintox) 3 hours

Fernos (pirimicarb) < 2 hours
fluvalinate (Mavrik, Spur) 2 hours

Gardona (tetrachlorvinphos) (lower rate) < 2 hours
Garrathion
Granulox EC (disulfoton) < 2 hours

heptachlor G (Velsicol) < 2 hours

isodrin
isolan (primin)
isopropyl-parathion < 2 hours

Korlan (ronnel) 1 day
Kroneton

Labaycid G or MA (fenthion)
Lannate LS (methomyl) 2 hours+
Larvin (thiodicarb) < 2 hours
Lorsban MA,ULV (chlorpyrifos) (0.045 lb/acre)

malathion ULV (3 fl oz/acre or less) 3 hours
Malonoben
Matacil ULV (aminocarb) (2.4 oz/acre or less) < 2 hours
Mavrik (fluvalinate) < 2 hours
menazon < 2 hours
Metasystox (demeton-S-methyl)
Metasystox-R (oxydemetonmethyl) < 2 hours
methoxychlor (DMDT, Marlate) 2 hours
MNFA (Nissol)
Mobilawn (dichlorfenthion) 2 hours
Morocide (binapacryl) < 2 hours

Nankor (fenchlorphos)
NDP (propyl thiopyrophosphate)
Neguvon (trichlorfon) 3-6 hours
Nemacide (dichlorfenthion) 2 hours
Niagra 9044 (binapacryl) < 2 hours
Nissol
Nogos MA (dichlorvos)
Nudrin LS (methomyl) 2 hours+
Nuvan MA (dichlorvos)

oil sprays (superior type) < 3 hours

Parsolin EC (disulfoton) 7 hours
Perthane (ethylan) 2 hours
phostex < 2 hours
Phosvel (leptophos) < 3 hours
Pirimor (pirimicarb) < 2 hours
Proxol (trichlorfon) 3-6 hours

Rabon (tetrachlorvinphos)
Rhothane (TDE) 2 hours
Ripcord (cypermethrin) (< 0.02 lb/acre) < 2 hours

Sapecron (chlorfenvinphos) < 2 hours
Saphi-Col, Sayfos (menazon) < 2 hours
Scout (tralomethrin) 2 hours
Sevin-4-oil (carbaryl) (0.5 lb/acre or less) 2 hours
Shirlan (sabadilla)
Solvigran, Solvirex EC (disulfoton) 7 hours
Spur (fluvalinate) 2 hours
Supona (chlorfenvinphos) < 2 hours
Syfos (menazon) < 2 hours
Systox (demeton) < 2 hours

TEPP < 5 hours
Thanite (isobornyl thiocyanate) < 3 hours
Thimet G (phorate) < 2 hours
Thiocron (amidithion)
Thiodan (endosulfan) (0.5 lb/acre or less) 2-3 hours
Tiguvon G.MA (fenthion)
Tiovel (endosulfan) (0.5 lb/acre or less) 2-3 hours
Torak (dalifor) < 2 hours
toxaphene (polychlorcamphene, Strobane) 2-4 hours
Tranid
Trigard (cyromazine) < 2 hours
Trithion (carbophenothion) 2-5 hours
Trolene (fenchlorphos)
Tugon (trichlorfon) 3-6 hours

Unden (propoxur) MA

Vapona ULV (dichlorvos) (0.1 lb/acre or less) < 2 hours
Vydate (oxamyl) (0.5 lb/acre or less) 3 hours

Wotexit (trichlorfon) 3-6 hours

Zolone (phosalone) 2 hours

+These materials are more hazardous to bees in a moist climate and under slow drying conditions.

GROUP D — CAN BE APPLIED AT ANY TIME WITH REASONABLE SAFETY TO BEES

Acaraben (chlorobenzilate)
Acaralate (chloropropylate)
Acarol (bromopropylate)
Akar (chlorobenzilate)
Akaritox (tetradifon)
allethrin
Altozar (hydroprene)
Ambush (permethrin)
Apollo (clofentezene)
azocyclotin

BAAM (amitraz)
Bacillus thuringiensis (Bactospeine, Bactur, Bakthane,
       Bug Time, Cekubacilina, Certan, Foil, Trident,
       Dipel, Sok-Bt)
Baygon G (propoxur)

chlorobenzilate
chloropropylate
Chlorparacide (chlorbenside)
Comite (propargite)
CPAS (chlorfensulphide)
CPBS (fenson)
CPCBS (chlorfenson)
Crotothane (dinocap)
Curater G (carbofuran)
Cryolite (fluoride)

Dasanit G (fensulfothion)
Dikar
Dimilin (diflubenzuron)
Dimite (chlorfenethol)
Di-Syston G (disulfoton)
Dithane (mancozeb, maneb, zineb)

DMC (chlorfenethol)
DN-111 or DNOCHP (dinex)

Folbex (chlorobenzilate)
Fundal (chlordimeform)
Furadan G (carbofuran)

Galecron (chlordimeform)
Genite 923 or Genitol 923
Granulox (disulfoton) G

Heliothis polyhedrosis virus (Elcar)

Karathane (dinocap)
Kelthane (dicofol)
Kepone (chlordecone)
Kroyocide (cryolite)

Largon (diflubenzuron)
Lethane 384 (butoxy thiocyanodiethyl ether)
lime sulfur
Lovozal (fenazaflor)

malathion G (Cythion, maldison,
mercaptothion)
margosan (neem oil)
Micasin (chlorfensulphide)
Milbex (chlorfensulphide-chlorfenethol)
Mirex G
Mitac (amitraz)
Mitox (chlorbenside)
Morestan (oxythioquinox)

Neoron (bromopropylate)
Neotran (oxythane)
nicotine sulfate

Oftanol (isofenphos)
Omite (propargite)
Ovex, Ovotran (chlorfenson)

Parsolin G (disulfoton)
PCPBS (fenson)
Pentac (dienochlor)
Plictran (cyhexatin)

Pounce (permethrin)
pyrethrum

Qikron (chlorfenethol)

Rospin (chloropropylate)
rotenone EC (Derris)
Ryanodine (ryania)

Savey (hexythiazox)
schradan (OMPA, Pestox III, Systam)
Sevin bait G (carbaryl)
Sevin G (carbaryl)
sodium fluosilicate baits
Solvigran or Solvirex G (disulfoton)
Sulphenone
sulfur

Tedion (tetradifon)
Terracur G (fensulfothion)
thiocyclam

Unden G (propoxur)

Vendex (fenbutatin-oxide)

Yaltox G (carbofuran)

Appendix III

TOXICITY OF INSECTICIDES AND MITICIDES TO WILD BEES

Length of residual toxic effects in hours or days

I. ALFALFA LEAFCUTTING BEE

GROUP A — DO NOT APPLY ON BLOOMING CROPS OR WEEDS

Advantage (carbosulfan) 2 days
Ambush (permethrin) > 2 days
Ammo (cypermethrin) (more than 0.025 lb/acre) > 3
     days
Avermectin (more than 0.025 lb/acre > 8 hours
Azodrin (monocrotophos) > 1 day

Baygon (propoxur) 1 day
Bidrin (dicrotophos) 2.5 days
Brigade (bifenthrin) > 1 day

Capture (bifenthrin) > 1 day
Carzol (formetanate) (0.5 lb/acre or more) 14 hours
Cidial (phenthoate) > 1 day
Cygon (dimethoate) > 3 days
Cymbush (cypermethrin) > 3 days

De-Fend (dimethoate) > 3 days

Di-Syston (disulfoton) 13 hours
diazinon 2 days
endrin 2.5 days

Furadan F (carbofuran) 7-14 days

Gardona D (tetrachlorvinphos) 1 day
Guthion (azinphosmethyl) > 3 days

Imidan (phosmet) > 1 day
Karate (cyhalothrin) > 1 day

Lannate (methomyl) (0.5 lb/acre or more) 6-15 hours
Lorsban (chlorpyrifos) 6-7 days

Malathion ULV 7 days
malathion > 2 days
methoxychlor D > 1 day
methyl parathion 1 day
Monitor (methamidophos) 1 day

Nudrin (methomyl) (0.5 lb/acre or more) 6-15 hours

Orthene (acephate) > 3 days

parathion 1-3 days DDT 1-2 days
Penncap-M (methyl parathion) > 8 days
phosphamidon 2 days
Pounce (permethrin) > 2 days
Primicid (pirimiphos-ethyl) > 2 days

Pydrin (fenvalerate) 9 hours
Rebelate (dimethoate) > 3 days

Sevin (carbaryl) 3-7 days
Supracide (methidathion) 1-3 days

Temik G (aldicarb) (applied at least 4 weeks before
        bloom)
Thiodan (endosulfan) (0.5 lb/acre) 1-3 days
Tiovel (endosulfan) (0.5 lb/acre) 1-3 days
toxaphene 3-7 days
Trithion (carbophenothion) 2 days

Vydate (oxamyl) (1 lb/acre) > 1 day

GROUP B — APPLY ONLY DURING LATE EVENING

Actellic (pirimiphos-methyl) 9 hours
Carzol (formetanate) (0.25 lb/acre or less) 4-12 hours
Delnav (dioxathion) 9 hours
Dibrom (naled) 12 hours
Lannate (methomyl) (0.25 lb/acre or less) < 4 hours
Larvin 8 hours
Metasystox-R (oxydemetonmethyl) < 4 hours
methoxychlor WP 12 hours
Nudrin (methomyl) (0.25 lb/acre or less) < 4 hours
Phosdrin (mevinphos) < 5 hours
Scout (tralomethrin) < 8 hours
TEPP < 5 hours
Vydate (oxamyl) (0.5 lb/acre or less) 3-9 hours

GROUP C — APPLY ONLY DURING LATE EVENING, NIGHT, OR EARLY MORNING.

Ammo (cypermethrin) (0.025 lb/acre or less) 2 hours
Avermectin (0.025 lb/acre or less) < 2 hours
Dylox (trichlorfon) 2-5 hours
Mavrik (fluvalinate) 2 hours
methoxychlor EC 2-4 hours
Pirimor (pirimicarb) < 2 hours
Proxol (trichlorfon) 2-5 hours
Sevin XLR (carbaryl) (1.5 lb/acres or less) 3 hours
Spur (fluvalinate) 2 hours
Systox (demeton) < 3 hours
Tedion (tetradifon) < 3 hours
Thimet G (phorate) < 2 hours
Zolone (phosalone) < 2 hours

GROUP D — CAN BE APPLIED AT ANY TIME WITH REASONABLE SAFETY TO BEES.

Baygon G (propoxur)
Comite (propargite)
Di-Syston G (disulfoton)
Furadan G (carbofuran)

Kelthane (dicofol)
Omite (propargite)

II. ALKALI BEE

GROUP A — DO NOT APPLY ON BLOOMING CROPS OR WEEDS.

Ambush (permethrin) 1-2 days
Azodrin (monocrotophos) > 1 day
Baygon (propoxur) 1 day
Bidrin (dicrotophos) 2-4 days
Carzol (1 lb/acre or more) > 1 day
Cidial (phenthoate) > 1 day
Cygon (dimethoate) 2-3 days
diazinon 1.5 days
dieldrin > 1 day
endrin 1 day
EPN 1 day
Furadan F (carbofuran) 7-14 days
Gardona D (tetrachlorvinphos) 1 day
Guthion (azinphosmethyl) 3 days
Imidan (phosmet) > 1 day
Lannate (methomyl) (0.5 lb/acre or more) > 1 day
Lorsban (chlorpyrifos) 3-6 days
malathion ULV > 5 days
malathion lb/acre 1 day
methyl parathion 1 day
Monitor (methamidophos) 1 day
Nudrin (methomyl) (0.5 lb/acre or more) > 1 day
Orthene (acephate) > 3 days
parathion 1 day
Penncap-M (methyl parathion) > 2 days
phosphamidon 1-2 days
Pounce (permethrin) 1-2 days
Pydrin (fenvalerate) (over 0.1
Rebelate (dimethoate) 2-3 days
Sevin (carbaryl) 3-7 days
Supracide (methidathion) 1-3 days > 1 day
Temik G (aldicarb) (applied at least 4 weeks before
　　　　　bloom)
Thiodan (endosulfan) (1 lb/acre or more) 1 day
Vydate (oxamyl) (1 lb/acre) > 1 day

GROUP B — APPLY ONLY DURING LATE EVENING

Carzol (formetanate) (1 lb/acre or more) 9 hours
Dibrom EC (naled) 12 hours
Dylox (trichlorfon) 6-14 hours
endrin 8 hours
Gardona EC (tetrachlorvinphos) 10 hours
Imidan (phosmet) (0.5 lb/acre or less) 12 hours
Lannate (methomyl) (0.5 lb/acre) 5-8 hours
Larvin (thiodicarb) 5 hours
malathion EC 1.5 days
Mavrik (fluvalinate) 8 hours
methoxychlor WP 8 hours
Nudrin (methomyl) (0.5 lb/acre) 5-8 hours
Phosdrin (mevinphos) < 5 hours
Proxol (trichlorfon) 6-14 hours
Pydrin (fenvalerate) (0.1 lb/acre or less) 8 hours
Scout (tralomethrin) < 8 hours
Spur (fluvalinate) 8 hours
Thiodan (endosulfan) (0.5 lb/acre or less) 5 hours
toxaphene 10 hours
Trithion (carbophenothion) 9 hours
Vydate (oxamyl) (0.5 lb/acre or less) 3-9 hours

GROUP C — APPLY ONLY DURING LATE EVENING, NIGHT, OR EARLY MORNING.

Avermectin < 2 hours
Carzol (formetanate) (0.5 lb/acre or less) 3 hours
DDT < 4 hours
Di-Syston EC (disulfoton) < 2 hours
Lannate (methomyl) (0.25 lb/acre or less) 2 hours
Metasystox-R (oxdemetonmethyl) < 2 hours
methoxychlor EC < 2 hours
Nudrin (methomyl) (0.25 lb/acre or less) 2 hours
Pirimor (pirimicarb) < 2 hours
schradan < 2 hours
Systox (demeton) < 2 hours
TEPP < 5 hours
Thimet G (phorate) < 2 hours
Zolone (phosalone) < 2 hours

GROUP D — CAN BE APPLIED AT ANY TIME WITH REASONABLE SAFETY TO BEES.

Baygon G (propoxur)
Comite (propargite)
Di-Syston G (disulfoton)
Furadan G (carbofuran)
Kelthane (dicofol)
Omite (propargite)
Tedion (tetradifon)

III. BUMBLE BEES

GROUP A — DO NOT APPLY ON BLOOMING CROPS OR WEEDS.

Bidrin (dicrotophos)
Cygon (dimethoate)
diazinon
Furadan F (carbofuran)
Guthion (azinphosmethyl)
Lorsban (chlorpyrifos)
malathion ULV
Orthene (acephate)
parathion
Penncap-M (methyl parathion)
Pydrin (fenvalerate)
Rebelate (dimethoate)
Sevin (carbaryl)
Supracide (methidathion)
Temik G (aldicarb) (applied at least 4 weeks before bloom)

GROUP B — APPLY ONLY DURING LATE EVENING.

Dibron (naled)
malathion EC
TEPP

GROUP C — APPLY ONLY DURING LATE EVENING, NIGHT, OR EARLY MORNING

Carzol (formetanate)
DDT
Di-Syston EC (disulfoton)
Dylox (trichlorfon)
Lannate (methomyl)
Metasystox-R (oxydemetonmethyl)
Nudrin (methomyl)
Systox (demeton)
toxaphene

GROUP D — CAN BE APPLIED AT ANY TIME WITH REASONABLE SAFETY TO BEES.

Baygon G (propoxur)
Comite (propargite)
Di-Syston G (disulfoton)
Furadan G (carbofuran)
Kelthane (dicofol)
Omite (propargite)

APPENDIX IV

TOXICITY OF HERBICIDES, BLOSSOM AND FRUIT THINNERS, DESICCANTS, AND PLANT GROWTH REGULATORS TO HONEY BEES

GROUP A — DO NOT APPLY ON BLOOMING CROPS OR WEEDS.

arsenic trioxide and other inorganic arsenicals
DNBP (dinoseb) (1.5 qt/100 gal or more)
Elgetol (dinitrocresol)
Sevin WP (carbaryl)

GROUP B — APPLY ONLY DURING LATE EVENING, NIGHT, OR EARLY MORNING ON BLOOMING CROPS OR WEEDS.

2,4-D (alkanolamine salts)
2,4-D (butoxyethanol ester)*
2,4-D (isopropyl ester)
Amino Triazole (amitrole)
Elgetol (dinitrocresol) (1.5 pt/100 gal or less)
endothall
Fusilade (fluazifop-butyl)
Hyvar X (bromacil)
Savit (carbaryl)
Sevin XLR (carbaryl)
Simazine
Weedone LV4 (butoxyethanol ester of 2,4-D)*

GROUP C — CAN BE APPLIED AT ANY TIME WITH REASONABLE SAFETY TO BEES.

2,4-D (butyl ether ester)*
2,4-D (sodium salts)
2,4-DB
2,4,5-T
Alar (daminozide)
Amiben (chloramben)
Ammate (AMS)
atrazine
Avenge (difenzoquat)
Banvel (dicamba)
Butoxone (2,4-DB)
Carbyne (barban)
Chloro IPC (chlorpropham)
dalapon
Desiccant (arsenic acid)
diquat
Eptam (EPTC)
Ethrel (ethephon)
Goal (oxyfluorfen)
IPC (propham)
Karmex (diuron)
Kerb (pronamide)
Lasso (alachlor)
MCPA
Monobor-chlorate
NAA (naphthaleneacetic acid)
paraquat 2,4-D (isooctyl ester)
Roundup (glyphosate)
Sencor (metribuzin)
Silvex (2,4,5-TP)
Sinbar (terbacil)
Tordon (picloram)
Treflan (trifluralin)

*There is field evidence that butyl derivatives of 2,4-D have a long-term chronic toxicity to bees, especially in cool climates and when nectar forage plants are treated.

APPENDIX V

TOXICITY OF FUNGICIDES TO HONEY BEES

GROUP A — APPLY ONLY DURING LATE EVENING, NIGHT, OR EARLY MORNING.

Morocide (binapacryl)

GROUP B — CAN BE APPLIED AT ANY TIME WITH REASONABLE SAFETY TO BEES.

Arasan (thiram)
Baycor (bitertanol)
Bayleton (triadimefon)
Benlate (benomyl)
Bordeaux mixture
captan*
copper sulfate
Cyprex (dodine)
Dessin (dinobuton)
Dikar (Dithane and Karathane)
Dithane M-22 (maneb)
Dithane M-45 (manzeb)
Dithane Z-78 (zineb)
ferbam
fixed copper
Funginex (triforine)
glyodin
Karathane (dinocap)
lime-sulfur
maneb
manzeb
Morestan (oxythioquinox)

Nustar
Phygon (dichlone)
prochoraz
Ronila (vinclozolin)
Roural (ipodione)
sulfur
Tag (PMA)
Thylate (thiram)
Vitavax (carboxin)
Zerlate (ziram)

* may cause larval mortality

APPENDIX VI

SPECIFIC BEE POISONING DATA FOR INSECTICIDES AND MITICIDES

(Including RT's, LD_{50}'s Dosage, etc.)

In the following data, pesticides are listed by established common name followed by trade names. Trade names are cross-referenced to the common name throughout. The uses are listed next: insecticide, miticide, fungicide, weedkiller, etc., and whether the chemical is systemic.

LD_{50}, the dose per individual bee which is expected to kill 50% of a group of bees based on laboratory studies, is given in micrograms per bee. All LD_{50} data were obtained from Atkins, E.L., D. Kellum and K.W. Atkins, *Reducing Pesticide Hazards to Honey Bees,* University of California, Division of Agricultural Science Leaflet 2883, 1981.

Highly toxic pesticides are those with an LD_{50} of 0.001 to 1.990 micrograms/bee; moderately toxic are those with an LD_{50} of 2.00 to 10.99 micrograms/bee; and relatively nontoxic are those with an LD_{50} of 11.00 or more micrograms/bee. The LD_{50} values (such as 0.175 micrograms/bee for parathion) are equivalent to the field dosage in pounds of active ingredient (AI) per acre that will kill 50% of the honey bees contacting the treated foliage or flying through the treated area during application of the pesticide. By using the slope of the mortality curve, they can predict the percent mortality of honey bees which will be caused by a given dosage in the field. The prediction will be reduced about 50% when the application is made at night and will be increased approximately 2 times when the application is made during maximum bee activity.

Since dead bees lose about 80% of their fresh weight in 2-4 days under field conditions, it is important to count the number of dead bees per sample before chemical analysis. The ppm figure is useful for relating to other chemical analyses of foliage, flowers, nectar, etc. It is derived by multiplying the LD_{50} by 7.8 (based on an average fresh weight per bee of 128 milligrams).

The second portion of the information is about **residual toxicity** and is based on our field tests conducted in the State of Washington in a relatively cool region. RT_{25} indicates the residual time required to reduce the activity of the chemical and bring bee mortality down to 25% in cage test exposures to field-weathered spray deposits. Materials with an RT_{25} of 2 hours or less can be applied with minimal hazard to bees when they are not actively foraging. Those reaching RT_{25} within 8 hours present a minimal problem to bees, if they are applied during late evening or night. Pest control programs sometimes require the use of an insecticide with an RT_{25} greater than 8 hours in order to adequately reduce potential damage of a valuable crop. Under such conditions, the RT_{40} is a more suitable criterion for determining the least hazardous chemical to use. Evening or night applications of a material with an RT_{40} of 8 hours or less are expected to cause only low to moderate kills of foraging honey bees. We have also included data for the alkali bee (AB) and the alfalfa leafcutting bee (LB). These pollinators are mainly of concern to alfalfa seed growers.

In the table the first number under the RT_{25}, RT_{40} column under each bee species refers to the RT_{25} and the second number, separated from the first by a ",", refers to the RT_{40}. If there is only one number that is the RT_{25}.

Formulations are: emulsifiable concentrate (EC), wettable powder (WP), dust (D), soluble powder (SP), solution (LS), and granular (G). Dosages are given in pounds of active ingredient (AI) per acre.

ACEPHATE   Insecticide
ORTHENE, ORTRAN

LD50-HB	MICROGRAM/BEE	PPM		
	1.2	9.4		
		RT_{25}, RT_{40}		
FORMULATION,DOSAGE		HB	AB	LB
SP, LS	0.5 LB AI/ACRE	1.5,1 DAYS	3,1.5 DAYS	3,1.5 DAYS
SP, LS	1.0 LB AI/ACRE	>3 DAYS	>3 DAYS	>3 DAYS

E.L. ATKINS GETS 2.5 DAYS OF RESIDUAL ACTION AGAINST HONEY BEES IN SOUTHERN CALIFORNIA.

ALDICARB   Insecticide, Miticide, Nematicide
TEMIK

LD50-HB	MICROGRAM/BEE	PPM		
	0.27	2.13		
		RT_{25}, RT_{40}		
FORMULATION,DOSAGE		HB	AB	LB
G	3 LB AI/ACRE			

NEVER CAUSES MORTALITY UNDER PROPER USE CONDITIONS
NOT HAZARDOUS TO BEES IN PACIFIC NORTHWEST WHEN INJECTED INTO SOIL AT LEAST 4 WEEKS BEFORE BLOOM AND THOROUGHLY IRRIGATED. OUR DATA SHOW THIS CHEMICAL WILL REPEL HONEY BEES ON CARROT SEED CROPS. E.L. ATKINS GETS NO RESIDUAL ACTION IN SOUTHERN CALIFORNIA.

ALDRIN   Insecticide

LD50-HB	MICROGRAM/BEE	PPM		
	0.35	2.75		
		RT_{25}, RT_{40}		
FORMULATION,DOSAGE		HB	AB	LB
EC	0.5 LB AI/ACRE	>1 DAY	NA	NA
EC	1.0 LB AI/ACRE	>2 DAYS	NA	NA

NO LONGER APPLIED TO ABOVE-GROUND PORTIONS OF CROPS.

ALLETHRIN   Insecticide
PYNAMIN

LD50-HB	MICROGRAM/BEE	PPM		
	NA	NA		
		RT_{25}, RT_{40}		
FORMULATION,DOSAGE		HB	AB	LB
EC	0.05 LB AI/ACRE	<2 HOURS	NA	NA

ALLETHRIN IS LOW IN HAZARD TO HONEY BEES.

FOR EXPLANATION OF DATA, SEE PAGES 160-161.

AMINOCARB Insecticide
MATACIL

LD50-HB	MICROGRAM/BEE	PPM
	1.12	8.75

RT_{25}, RT_{40}

FORMULATION,DOSAGE		HB	AB	LB
F	1.0 LB AI/ACRE	>3 DAYS	NA	NA
ULV	0.26 LB AI/ACRE	<2 HOURS	NA	NA

ULV APPLICATION RELATIVELY LOW IN HAZARD TO HONEY BEES AT 0.26 LB AI / ACRE.

AMITRAZ Insecticide, Miticide
MITAC, BAAM, ECTODEX

LD50-HB	MICROGRAM/BEE	PPM
	NA	NA

RT_{25}, RT_{40}

FORMULATION,DOSAGE		HB	AB	LB
EC	0.5-1.0 LB AI/ACRE	<2 HOURS	NA	NA
EC	1.5-2.0 LB AI/ACRE	2 HOURS	NA	NA

AMITRAZ IS LOW IN HAZARD TO HONEY BEES.

AVERMECTIN Insecticide

LD50-HB	MICROGRAM/BEE	PPM
	NA	NA

RT_{25}, RT_{40}

FORMULATION,DOSAGE		HB	AB	LB
LS	0.0125 LB AI/ACRE	6 HOURS	<2 HOURS	<2 HOURS
LS	0.025 LB AI/ACRE	8 HOURS	<2 HOURS	<2 HOURS
LS	0.05 LB AI/ACRE	1 DAY	<2 HOURS	<2 HOURS
LS	0.1 LB AI/ACRE	3 DAYS	<2 HOURS	6 HOURS

AVERMECTIN IS MODERATELY HAZARDOUS TO HONEY BEES AT 0.025 LB AI /ACRE OR LESS, BUT IS LOW IN HAZARD TO ALKALI BEES AND ALFALFA LEAFCUTTING BEES.

AZINPHOSMETHYL Insecticide, Miticide
GUTHION, GUSATHION

LD50-HB	MICROGRAM/BEE	PPM
	0.43	3.34

RT_{25}, RT_{40}

FORMULATION,DOSAGE		HB	AB	LB
EC	0.375 LB AI/ACRE	2.5,2 DAYS	3 DAYS	3,2.5 DAYS
WP	0.375 LB AI/ACRE	4 DAYS	NA	NA
WP	0.8 LB AI/ACRE	>5 DAYS	NA	NA

E.L. ATKINS GETS 5 DAYS OF RESIDUAL ACTION AGAINST HONEY BEES IN SOUTHERN CALIFORNIA.

AZOCYCLOTIN   Miticide
PEROPAL, CLERMAIT

LD50-HB	MICROGRAM/BEE	PPM
	NA	NA

FORMULATION,DOSAGE		RT_{25}, RT_{40}	
		HB	AB	LB
WP	0.75 LB AI/ACRE	<2 HOURS	NA	NA
WP	0.5 LB AI/ACRE	<2 HOURS	NA	NA
WP	0.25 LB AI/ACRE	<2 HOURS	NA	NA

ESSENTIALLY NONTOXIC TO HONEY BEES.

BACILLUS THURINGIENSIS   Microbial Insecticide
BACTUR, BACTOSPEINE, THURICIDE, CERTAN, DIPEL, FOIL, TRIDENT

LD50-HB	MICROGRAM/BEE	PPM
	NA	NA

FORMULATION,DOSAGE		RT_{25}, RT_{40}		
		HB	AB	LB
WP, F	1.0 LB PRODUCT/ACRE	<2 HOURS	<2 HRS.	<2 HRS.

ESSENTIALLY NONTOXIC TO HONEY BEES.

BENDIOCARB   Insecticide
FICAM, TURCAM

LD50-HB	MICROGRAM/BEE	PPM
	0.43	3.34

FORMULATION,DOSAGE		RT_{25}, RT_{40}		
		HB	AB	LB
D	1-2 LB AI/ACRE	>1 DAY	NA	NA
WP	1-2 LB AI/ACRE	>1 DAY	NA	NA

BENDIOCARB IS HIGHLY HAZARDOUS TO HONEY BEES. MAINLY USED AGAINST HOUSEHOLD PESTS AND ON ORNAMENTALS.

BINAPACRYL   Miticide
MOROCIDE, ACRICID, ENDOSAN

LD50-HB	MICROGRAM/BEE	PPM
	NA	NA

FORMULATION,DOSAGE		RT_{25}, RT_{40}		
		HB	AB	LB
WP	0.5 LB AI/ACRE	<2 HOURS	NA	NA

ESSENTIALLY NONTOXIC TO HONEY BEES.

FOR EXPLANATION OF DATA, SEE PAGES 160-161.

BOMYL   Insecticide, Miticide
SWAT

LD50-HB	MICROGRAM/BEE	PPM		
	NA	NA		

		RT_{25}, RT_{40}		
FORMULATION,DOSAGE		HB	AB	LB
EC	0.5 LB AI/ACRE	2 DAYS	NA	>1 DAY

MAINLY USED IN A FLY BAIT.

BROMOPHOS   Insecticide
NEXION

LD50-HB	MICROGRAM/BEE	PPM
	NA	NA

		RT_{25}, RT_{40}		
FORMULATION,DOSAGE		HB	AB	LB
EC	1.0 LB AI/ACRE	14,7 HOURS	NA	>1 DAY
EC	0.25 LB AI/ACRE	<3 HOURS	NA	>1 DAY

BROMOPHOS IS LOW TO MODERATE IN HAZARD TO HONEY BEES, BUT HIGHLY HAZARDOUS TO ALFALFA LEAFCUTTING BEES.

BROMOPHOS-ETHYL   Insecticide, Miticide
NEXAGAN

LD50-HB	MICROGRAM/BEE	PPM
	NA	NA

		RT_{25}, RT_{40}		
FORMULATION,DOSAGE		HB	AB	LB
EC	1.0 LB AI/ACRE	>1 DAY	NA	NA

CARBARYL   Insecticide, Fruit Thinner
SEVIN, SAVIT

LD50-HB	MICROGRAM/BEE	PPM
	1.5	12.0
SEVIN SL		
	13.7	107.2
SEVIN XLR		
	26.5	207.3

		RT_{25}, RT_{40}		
FORMULATION,DOSAGE		HB	AB	LB
WP	2.0 LB AI/ACRE	7-12 DAYS	NA	NA
WP	1.0 LB AI/ACRE	3-7 DAYS	3-7 DAYS	3-7 DAYS
WP	0.5 LB AI/ACRE	20,16 HOURS	NA	NA
F SEVN4OIL	1.0 LB AI/ACRE	>3 DAYS	>3 DAYS	>3,2.5 DAYS
F SEVN4OIL	0.5 LB AI/ACRE	2,<2 HOURS	NA	3,<2 HOURS
F SEVINXLR	2.0 LB AI/ACRE	2,1 DAYS	NA	NA
F SEVINXLR	1.5 LB AI/ACRE	20,8 HOURS	NA	NA
F SEVINXLR	1.0 LB AI/ACRE	<2 HOURS	NA	<2 HOURS
F SEVIMOL	1.0 LB AI/ACRE	3,1 DAYS	NA	NA
WP+MOLASSES	1.0 LB AI+1 GAL/ACRE	24,17hr	NA	NA

FOR EXPLANATION OF DATA, SEE PAGES 160-161.

D	2.0 LB AI/ACRE	3-14 DAYS	NA	NA
G,BAIT	2.0 LB AI/ACRE	<2 HOURS	NA	NA

GRANULAR FORMULATIONS AND BAITS WITH CARBARYL ARE ESSENTIALLY NONTOXIC TO HONEY BEES. E.L. ATKINS GETS 3-7 DAYS RESIDUAL TOXICITY OF CARBARYL TO HONEY BEES IN SOUTHERN CALIFORNIA. SEVIN XLR AT 1.5 LB AI / ACRE OR LESS IS LOW IN HAZARD TO HONEY BEES IN ARID AREA TESTING, BUT IS REPORTED TO BE MORE HAZARDOUS UNDER HUMID CONDITIONS.

CARBOFURAN Insecticide, Systemic
FURADAN,CURATERR

LD50-HB	MICROGRAM/BEE	PPM
	0.15	1.16

RT_{25}, RT_{40}

FORMULATION,DOSAGE		HB	AB	LB
F	1.0 LB AI/ACRE	7->14 DAYS	7->14 DAYS	7->14 DAYS
F	0.5 LB AI/ACRE	>1 DAY	NA	NA
G	2.0 LB AI/ACRE	<2 HOURS	<2 HOURS	<2 HOURS

E.L. ATKINS GETS 3 TO >5 DAYS RESIDUAL ACTION IN SOUTHERN CALIFORNIA.

CARBOPHENOTHION Insecticide, Miticide
TRITHION

LD50-HB	MICROGRAM/BEE	PPM
	12.99	101.48

RT_{25}, RT_{40}

FORMULATION,DOSAGE		HB	AB	LB
F	1.0 LB AI/ACRE	<2-5,<2 HOURS	9,5 HOURS	2,1.5 DAYS
D	1.0 LB AI/ACRE	>1 DAY	NA	NA

MODERATELY TOXIC TO HONEY BEES.

CARBOSULFAN Insecticide, Nematicide, Miticide
ADVANTAGE, MARSHAL, POSSE

LD50-HB	MICROGRAM/BEE	PPM
	0.68	5.30

RT_{25}, RT_{40}

FORMULATION,DOSAGE		HB	AB	LB
EC	0.5 LB AI/ACRE	>3 DAYS	NA	2,1.5 DAYS
EC	0.25 LB AI/ACRE	>3 DAYS	NA	1.5,1 DAYS

E.L. ATKINS GETS 3.5 DAYS RESIDUAL ACTION IN SOUTHERN CALIFORNIA.

CHLORDANE Insecticide
OCTACHLOR

LD50-HB	MICROGRAM/BEE	PPM
	8.8	68.75

RT_{25}, RT_{40}

FORMULATION,DOSAGE		HB	AB	LB
EC	1-2 LB AI/ACRE	<2 HOURS	NA	NA

CHLORDANE USED FOR ANT CONTROL HAS AN AFFINITY FOR BEESWAX AND WILL KILL BEES WHENEVER CONTAMINATED COMBS ARE USED

FOR EXPLANATION OF DATA, SEE PAGES 160-161.

CHLORDIMEFORM   Insecticide, Miticide
GALECRON, FUNDAL

LD50-HB MICROGRAM/BEE PPM
             NA   NA

		RT_{25}, RT_{40}	
FORMULATION,DOSAGE	HB	AB	LB
SP,EC 1.0 LB AI/ACRE	<2 HOURS	<2 HOURS	<2 HOURS
D 1.0 LB AI/ACRE	<3 HOURS	NA	NA
EC 0.5 LB AI/ACRE	<2 HOURS	NA	NA

RELATIVELY NONTOXIC TO BEES.

CHLORFENSULFIDE   Miticide
MICASIN, MILBEX

LD50-HB MICROGRAM/BEE PPM
                NA   NA

		RT_{25}, RT_{40}	
FORMULATION,DOSAGE	HB	AB	LB
WP 1.0 LB AI/ACRE	<2 HOURS	NA	NA

RELATIVELY NONTOXIC TO BEES.

CHLOROPROPYLATE   Miticide
ACARALATE , ROSPIN

LD50-HB MICROGRAM/BEE PPM
             NA   NA

		RT_{25}, RT_{40}	
FORMULATION,DOSAGE	HB	AB	LB
WP, EC 0.5 LB AI/ACRE	<2 HOURS	NA	NA

CHLORPYRIFOS   Insecticide, Miticide
LORSBAN, DURSBAN

LD50-HB MICROGRAM/BEE PPM
              0.11   0.86

		RT_{25}, RT_{40}	
FORMULATION,DOSAGE	HB	AB	LB
EC 1.0 LB AI/ACRE	6,5.5 DAYS	6,5.5 DAYS	7,6 DAYS
EC 0.5 LB AI/ACRE	4,3 DAYS	3,2.5 DAYS	6,5.5 DAYS
EC 0.25 LB AI/ACRE	18,9 HOURS	21,12 HOURS	2,1.5 DAYS
EC 0.05 LB AI/ACRE	8,5 HOURS	NA	NA

CAN CAUSE ABOUT 50% REDUCED VISITATION FOR UP TO 7 DAYS, BUT NOT
COMPLETELY ENOUGH TO SAFEGUARD BEES. RELATIVELY SAFE TO BEES WHEN
APPLIED AT NIGHT AT THE MOSQUITO ABATEMENT RATE, 0.05 LB AI/ACRE. E.L.
ATKINS GETS 2 TO 3.5 DAYS OF RESIDUAL ACTION IN SOUTHERN CALIFORNIA.

CRYOLITE   Insecticide
KRYOCIDE

LD50-HB	MICROGRAM/BEE	PPM
	NA	NA

RT_{25}, RT_{40}

FORMULATION,DOSAGE	HB	AB	LB
DUST 48-72 LB AI/ACRE	<2 HOURS	NA	<2 HOURS

RELATIVELY NONTOXIC TO BEES.

CYHEXATIN   Miticide
PLICTRAN

LD50-HB	MICROGRAM/BEE	PPM
	NA	NA

RT_{25}, RT_{40}

FORMULATION,DOSAGE	HB	AB	LB
WP 0.5 LB AI/A	<2 HOURS	NA	NA

RELATIVELY NONTOXIC TO BEES.

CYPERMETHRIN   Insecticide
AMMO,CYMBUSH

LD50-HB	MICROGRAM/BEE	PPM
	NA	NA

RT_{25}, RT_{40}

FORMULATION,DOSAGE	HB	AB	LB
EC 0.075 LB AI/ACRE	>3 DAYS	NA	>3 DAYS
EC 0.025 LB AI/ACRE	<2 HOURS	NA	<2 HOURS

LITTLE OR NO HAZARD TO BEES WHEN APPLIED LATE EVENING ,NIGHT OR EARLY MORNING AT THE 0.025 LB AI/ACRE DOSAGE.

DDT   Insecticide

LD50-HB	MICROGRAM/BEE	PPM
	6.19	48.36

RT_{25}, RT_{40}

FORMULATION,DOSAGE	HB	AB	LB
EC 2.0 LB AI/ACRE	4-42,2-8 HOURS	6-24,2-6 HOURS	1-2,0.4-1 DAYS
D 2.0 LB AI/ACRE	2-3 DAYS	NA	NA

RELATIVELY LOW HAZARD TO HONEY BEES AND ALKALI BEES WHEN APPLIED AS A SPRAY DURING LATE EVENING, NIGHT OR EARLY MORNING.

DEMETON   Insecticide, Miticide, Systemic
SYSTOX

LD50-HB	MICROGRAM/BEE	PPM
	1.7	13.36

RT_{25}, RT_{40}

FORMULATION,DOSAGE	HB	AB	LB
EC 0.25 LB AI/ACRE	<2 HOURS	<2 HOURS	<2 HOURS
EC 2LB/GAL0.375 LB AI/ACRE	<2 HOURS	NA	<2, <2 HOURS

168

EC 6LB/GAL0.375 LB AI/ACRE <2 HOURS   NA   6,<2 HOURS

THE 6 LB/GAL FORMULATION OF DEMETON IS SOMEWHAT MORE TOXIC TO ALFALFA LEAFCUTTING BEES THAN THE 2 LB/GAL FORMULATION. THIS DIFFERENCE IS NOT DETECTABLE WITH THE HONEY BEE.

DIALIFOR   Insecticide, Miticide
TORAK

LD50-HB	MICROGRAM/BEE	PPM		
	NA	NA		
			RT_{25}, RT_{40}	
FORMULATION,DOSAGE		HB	AB	LB
EC 1.0 LB AI/100 GAL		<2 HOURS	NA	NA

RELATIVELY LOW HAZARD TO HONEY BEES.

DIAZINON   Insecticide
SPECTRACIDE

LD50-HB	MICROGRAM/BEE	PPM		
	0.37	2.9		
			RT_{25}, RT_{40}	
FORMULATION,DOSAGE		HB	AB	LB
EC 1.0 LB AI/ACRE		2,1.7 DAYS	1.5,1 DAYS	2,1.5 DAYS
EC 0.5 LB AI/ACRE		<1 DAY	NA	NA

E.L. ATKINS GETS 1 TO 2 DAYS RESIDUAL ACTION IN SOUTHERN CALIFORNIA.

DICHLOFENTHION   Insecticide, Nematicide
NEMACIDE, MOBILAWN

LD50-HB	MICROGRAM/BEE	PPM		
	NA	NA		
			RT_{25}, RT_{40}	
FORMULATION,DOSAGE		HB	AB	LB
EC 1.3 LB AI/ACRE		<2 HOURS	NA	NA

RELATIVELY LOW HAZARD TO HONEY BEES.

DICHLORVOS   Insecticide
VAPONA,DDVP

LD50-HB	MICROGRAM/BEE	PPM		
	0.5	3.9		
			RT_{25}, RT_{40}	
FORMULATION,DOSAGE		HB	AB	LB
EC 1.0 LB AI/ACRE		>1 DAY	NA	NA
ULV 0.1LBAI/ACRE		<2 HOURS	NA	NA

RELATIVELY LOW HAZARD TO HONEY BEES WHEN APLIED AT NIGHT AT MOSQUITO ABATEMENT RATE, 0.1 LB AI/ACRE.

DICOFOL   Miticide
KELTHANE

LD50-HB	MICROGRAM/BEE	PPM
	NA	NA

		RT_{25}, RT_{40}	
FORMULATION,DOSAGE		HB	AB	LB
EC	1.0 LB AI/ACRE	<2 HOURS	<2 HOURS	<2 HOURS

ESSENTIALLY NONTOXIC TO BEES.

DICROTOPHOS   Insecticide, Systemic
BIDRIN

LD50-HB	MICROGRAM/BEE	PPM
	0.3	2.38

		RT_{25}, RT_{40}	
FORMULATION,DOSAGE		HB	AB	LB
EC	0.5 LB AI/ACRE	1-1.5 DAYS	2-3 DAYS	1-3 DAYS

E.L. ATKINS GETS 2 TO 4 DAYS RESIDUAL ACTION IN SOUTHERN CALIFORNIA.

DIELDRIN   Insecticide

LD50-HB	MICROGRAM/BEE	PPM
	0.13	1.04

		RT_{25}, RT_{40}	
FORMULATION,DOSAGE		HB	AB	LB
D	0.5 LB AI/ACRE	8 DAYS	NA	NA
WP	0.5 LB AI/ACRE	5-7 DAYS	NA	NA
EC	0.5 LB AI/ACRE	<2 DAYS	NA	NA
G	2.0 LB AI/ACRE	<2 HOURS	NA	NA

E.L. ATKINS GETS 1.5 TO 5 DAYS RESIDUAL ACTION IN SOUTHERN CALIFORNIA.

DIENOCHLOR   Miticide
PENTAC

LD50-HB	MICROGRAM/BEE	PPM
	NA	NA

		RT_{25}, RT_{40}	
FORMULATION,DOSAGE		HB	AB	LB
EC	1.0 LB AI/ACRE	<2 HOURS	NA	NA

RELATIVELY NONTOXIC TO BEES.

DIFLUBENZURON   Insecticide
DIMILIN

LD50-HB	MICROGRAM/BEE	PPM
	NA	NA

FOR EXPLANATION OF DATA, SEE PAGES 160-161.

FORMULATION,DOSAGE	RT_{25}, RT_{40} HB	AB	LB
NA 0.125 LB AI/ACRE	<2-6 HOURS	<2-6 HOURS	<2-6 HOURS

WAS NOT HAZARDOUS TO HONEY BEE BROOD IN LARGE-SCALE FIELD TESTS AT UP TO 0.25 LB AI/ACRE.

DIKAR Fungicide, Miticide

LD50-HB MICROGRAM/BEE	PPM
NA	NA

FORMULATION,DOSAGE	RT_{25}, RT_{40} HB	AB	LB
WP 1.5 LB AI/100 GAL	<2 HOURS	NA	NA

ESSENTIALLY NONTOXIC TO HONEY BEES.

DIMETHOATE   Insecticide, Miticide, Systemic
CYGON, DE-FEND, REBELATE, ROGOR

LD50-HB MICROGRAM/BEE	PPM
0.19	1.49

FORMULATION,DOSAGE	RT_{25}, RT_{40} HB	AB	LB
EC 0.5 LB AI/ACRE	0.4->3 DAYS	2->3 DAYS	3->3 DAYS
EC 0.25 LB AI/ACRE	4,<2 HOURS	NA	<2 HOURS
EC 0.125 LB AI/ACRE	<2 HOURS	NA	<2 HOURS
G 10 LB AI/ACRE	NA	NA	14.5,9 DAYS
WP 0.5 LB AI/ACRE	>1 DAY	NA	NA

THERE IS A DRAMATIC REDUCTION IN MORTALITY OF BEES WHEN REDUCED DOSAGES ARE APPLIED IN SMALL-SCALE STUDIES. WE HAVE NO FIELD EXPERIENCE TO SHOW WHETHER THIS WILL HOLD FOR DIMETHOATE IN LARGE-SCALE OPERATIONS. E.L. ATKINS GETS 1 TO 3.5 DAYS RESIDUAL ACTION IN SOUTHERN CALIFORNIA.OTHER INVESTIGATORS HAVE FOUND BOTH REPELLENT AND TOXICITY PROBLEMS FOR HONEY BEES WITH DIMETHOATE APPLIED TO LEMONS OR ONION SEED CROPS IN THE SOUTHWEST.

DINITROCRESOL   Weedkiller, Insecticide, Desiccant, Blossom Thinner
ELGETOL, SINOX

LD50-HB MICROGRAM/BEE	PPM
NA	NA

FORMULATION,DOSAGE	RT_{25}, RT_{40} HB	AB	LB
F 1.5QT/100GAL OR MORE	>1 DAY	NA	NA
F 1.5PT/100GAL OR LESS	2 HOURS	NA	NA

DINITROCRESOL USED AS A BLOSSOM THINNER IN APPLE ORCHARDS HAS NOT BEEN HAZARDOUS TO HONEY BEES AT 1.5 PT/100 GAL OR LESS.

DINOCAP   Insecticide, Fungicide
KARATHANE

LD50-HB	MICROGRAM/BEE	PPM		
	NA	NA		
			RT_{25}, RT_{40}	
FORMULATION,DOSAGE		HB	AB	LB
WP	0.75 LB AI/ACRE	<2 HOURS	NA	NA

ESSENTIALLY NONTOXIC TO HONEY BEES.

DIOXATHION   Insecticide, Miticide
DELNAV, NAVADEL

LD50-HB	MICROGRAM/BEE	PPM		
	NA	NA		
			RT_{25}, RT_{40}	
FORMULATION,DOSAGE		HB	AB	LB
EC	1.0 LB AI/ACRE	<2	NA	9,<2

RELATIVELY NONTOXIC TO BEES.

DISULFOTON   Insecticide, Systemic, Miticide
DI-SYSTON, SOLVIREX

LD50-HB	MICROGRAM/BEE	PPM		
	6.12	47.8		
			RT_{25}, RT_{40}	
FORMULATION,DOSAGE		HB	AB	LB
EC	0.5 LB AI/ACRE	<2 HOURS	<2 HOURS	13,6 HOURS
EC	1.0 LB AI/ACRE	7,4 HOURS	3,<2 HOURS	22,16 HOURS
G	1.0 LB AI/ACRE	<2 HOURS	<2 HOURS	<2 HOURS

DISULFOTON GRANULES ARE NONHAZARDOUS TO BEES.

ENDOSULFAN   Insecticide, Miticide
TIOVEL, THIODAN, CYCLODAN

LD50-HB	MICROGRAM/BEE	PPM		
	7.8	61.0		
			RT_{25}, RT_{40}	
FORMULATION,DOSAGE		HB	AB	LB
EC	0.5 LB/ACRE	<2-3,<2 HOURS	5,<2 HOURS	1.4-3 DAYS
EC	1.0 LB/ACRE	5,3 HOURS	8 HOURS	>1 DAY
EC	1.5 LB/ACRE	7,5 HOURS	14 HOURS	>1 DAY
D	1.0 LB AI/ACRE	24,11 HOURS	>1 DAY	>1 DAY

ENDOSULFAN IS MODERATELY TOXIC TO HONEY BEES AND CAN BE USED WITH REASONABLE SAFETY DURING LATE EVENING, NIGHT OR EARLY MORNING.

FOR EXPLANATION OF DATA, SEE PAGES 160-161.

ENDRIN   Insecticide

LD50-HB	MICROGRAM/BEE	PPM
	2.04	15.94

RT_{25}, RT_{40}

FORMULATION,DOSAGE		HB	AB	LB
EC	0.5 LB AI/ACRE	<2 HOURS	2,0.5 DAYS	2.5,2 DAYS

ENDRIN IS RELATIVELY SAFE TO HONEY BEES WHEN APPLIED DURING LATE EVENING, NIGHT OR EARLY MORNING. IT IS MUCH MORE HAZARDOUS TO ALKALI BEES AND ALFALFA LEAFCUTTING BEES.

EPN   Insecticide

LD50-HB	MICROGRAM/BEE	PPM
	0.24	1.85

RT_{25}, RT_{40} 0

FORMULATION,DOSAGE		HB	AB	LB
WP	1.0 LB AI/ACRE	1 DAY	NA	NA

E.L. ATKINS GETS 1.5 TO 3 DAYS RESIDUAL ACTION IN SOUTHERN CALIFORNIA.

ETHION   Insecticide, Miticide
NIALATE

LD50-HB	MICROGRAM/BEE	PPM
	NA	NA

RT_{25}, RT_{40}

FORMULATION,DOSAGE		HB	AB	LB
EC	1.0 LB AI/ACRE	< 2 HOURS	NA	NA
WP	1.0 LB AI/ACRE	< 2 HOURS	NA	NA
G	1.0 LB AI/ACRE	< 2 HOURS	NA	NA

RELATIVELY NONTOXIC TO BEES.

ETHYLAN   Insecticide
PERTHANE

LD50-HB	MICROGRAM/BEE	PPM
	4.57	35.7

RT_{25}, RT_{40}

FORMULATION,DOSAGE		HB	AB	LB
EC	1.0 LB AI/ACRE	2 HOURS	NA	NA
D	1.0 LB AI/ACRE	>1 DAY	NA	NA

ETHYLAN IS MODERATELY TOXIC TO HONEY BEES.

ETRIMFOS   Insecticide
EKAMET

LD50-HB	MICROGRAM/BEE	PPM
	0.26	2.06

FOR EXPLANATION OF DATA, SEE PAGES 160-161.

		RT_{25}, RT_{40}		
FORMULATION,DOSAGE		HB	AB	LB
EC	0.5 LB AI/ACRE	4,3 DAYS	>5,3 DAYS	>5,3 DAYS
EC	0.25 LB AI/ACRE	>3,3 DAYS	4,3 DAYS	4,3 DAYS

ETRIMFOS SPRAYS CANNOT BE SAFELY APPLIED TO BLOOMING CROPS WHEN BEES ARE PRESENT.

FENAMIPHOS   Nematicide
NEMACUR

LD50-HB	MICROGRAM/BEE	PPM
	1.43	11.17

		RT_{25}, RT_{40}		
FORMULATION,DOSAGE		HB	AB	LB
EC	5.0 LB AI/ACRE	>1 DAY	NA	NA
G	5.0 LB AI/ACRE	<2 HOURS	NA	NA

FENBUTATIN-OXIDE   Miticide
VENDEX

LD50-HB	MICROGRAM/BEE	PPM
	NA	NA

		RT_{25}, RT_{40}		
FORMULATION,DOSAGE		HB	AB	LB
WP	0.125-0.5LBAI/100GAL	<2 HOURS	NA	NA

FENBUTATIN-OXIDE IS ESSENTIALLY NONTOXIC TO HONEY BEES.

FENITROTHION   Insecticide
SUMITHION

LD50-HB	MICROGRAM/BEE	PPM
	0.18	1.38

		RT_{25}, RT_{40}		
FORMULATION,DOSAGE		HB	AB	LB
EC	0.5 LB AI/ACRE	24,16 HOURS	3,4 DAYS	3,5 DAYS
EC	1.0 LB AI/ACRE	4,3 DAYS	4.5,5 DAYS	>5 DAYS
WP	0.5 LB AI/ACRE	2,1.5 DAYS	NA	NA
WP	1.0 LB AI/ACRE	4,3 DAYS	NA	NA

FENITROTHION SPRAYS CANNOT BE SAFELY APPLIED TO BLOOMING CROPS WHEN BEES ARE PRESENT.

FENSULFOTHION   Insecticide
DASANIT

LD50-HB	MICROGRAM/BEE	PPM
	0.34	2.63

		RT_{25}, RT_{40}		
FORMULATION,DOSAGE		HB	AB	LB
EC	0.75 LB AI/ACRE	1 DAY	NA	NA
G	1.0 LB AI/ACRE	<2 HOURS	NA	NA

FOR EXPLANATION OF DATA, SEE PAGES 160-161.

FENSULFOTHION SPRAYS CANNOT BE SAFELY APPLIED TO BLOOMING CROPS WHEN BEES ARE PRESENT.

FENTHION   Insecticide
BAYTEX

LD50-HB   MICROGRAM/BEE   PPM
                0.32   2.49

FORMULATION,DOSAGE		HB	AB	LB
			RT_{25}, RT_{40}	
WP,EC	1.0 LB AI/ACRE	2-3 DAYS	NA	NA
ULV	0.1 LB AI/ACRE	<2 HOURS	NA	<2 HOURS
LC	0.05 LB AI/ACRE	<2 HOURS	NA	<2 HOURS

FENTHION AT THE MOSQUITO ABATEMENT RATE OF 0.1 LB AI/ACRE OR LESS APPLIED IN A QUART OF DIESEL OIL/ACRE IS MINIMALLY HAZARDOUS TO BEES.

FENVALERATE   Insecticide
PYDRIN

LD50-HB   MICROGRAM/BEE   PPM
                0.41   3.19

FORMULATIONLDOSAGE		HB	AB	LB
			RT_{25}, RT_{40}	
EC	0.1 LB AI/ACRE	6,5 HOURS	7,5 HOURS	9,7 HOURS
EC	0.4 LB AI/ACRE	>8 HOURS	>8 HOURS	>8 HOURS

E.L. ATKINS GETS 1 DAY OF RESIDUAL ACTION IN SOUTHERN CALIFORNIA. FENVALERATE IS SAFE FOR BEES BY ITS REPELLENT ACTION, AT LEAST IN ARID AREAS.

FLUVALINATE   Insecticide, Miticide
SPUR, MAVRIK, APISTAN

LD50-HB   MICROGRAM/BEE   PPM
                65.8   514.5

FORMULATION,DOSAGE		HB	AB	LB
			RT_{25}, RT_{40}	
EC	0.1 LB AI/ACRE	<2 HOURS	6,4 HOURS	<2 HOURS
EC	0.2 LB AI/ACRE	<2 HOURS	9,7 HOURS	<2 HOURS

FLUVALINATE IS ESSENTIALLY NONTOXIC TO BEES.

FONOFOS
DYFONATE

LD50-HB   MICROGRAM/BEE   PPM
                8.68   67.8

FORMULATION,DOSAGE		HB	AB	LB
			RT_{25}, RT_{40}	
EC	1.0 LB AI/ACRE	<2 HOURS	NA	NA

FOR EXPLANATION OF DATA, SEE PAGES 160-161.

EC	2.0 LB AI/ACRE	6,5 HOURS	NA	NA

FONOFOS IS MODERATELY TOXIC TO HONEY BEES.

FORMETANATE HYDROCHLORIDE   Insecticide, Miticide
CARZOL

LD50-HB	MICROGRAM/BEE	PPM
	9.2	71.95

RT_{25}, RT_{40}

FORMULATION,DOSAGE		HB	AB	LB
SP	0.92 LB AI/ACRE	<2-2 HOURS	3-9,<2-2 HOURS	14-20,7-14 HOURS
SP	0.46 LB AI/ACRE	<2 HOURS	3,<2 HOURS	3-4,<2 HOURS

FORMETANATE HYDROCHLORIDE IS MODERATELY TOXIC TO HONEY BEES. IT IS MUCH MORE HAZARDOUS TO ALFALFA LEAFCUTTING BEES.

***HELIOTHIS* POLYHEDROSIS VIRUS**   Microbial Insecticide
ELCAR

LD50-HB	MICROGRAM/BEE	PPM
	NA	NA

RT_{25}, RT_{40}

FORMULATION,DOSAGE		HB	AB	LB
WP	0.5 LB PRODUCT/ACRE	<2 HOURS	NA	NA

HELIOTHIS POLYHEDROSIS VIRUS IS ESSENTIALLY NONTOXIC TO HONEY BEES.

HEPTACHLOR   Insecticide

LD50-HB	MICROGRAM/BEE	PPM
	0.53	4.11

RT_{25}, RT_{40}

FORMULATION,DOSAGE		HB	AB	LB
EC	0.5 LB AI/ACRE	>1 DAY	NA	NA
G	2.0 LB AI/ACRE	>2 HOURS	NA	NA

HEPTACHLOR SPRAYS CANNOT BE SAFELY APPLIED TO BLOOMING CROPS WHEN BEES ARE PRESENT.

ISOBORNYL THIOCYANATE   Insecticide
THANITE

LD50-HB	MICROGRAM/BEE	PPM
	NA	NA

RT_{25}, RT_{40}

FORMULATION,DOSAGE		HB	AB	LB
EC	1.0 LB AI/ACRE	<3 HOURS	NA	NA

ISOBORNYL THIOCYANATE IS LOW IN HAZARD TO HONEY BEES.

FOR EXPLANATION OF DATA, SEE PAGES 160-161.

ISOFENPHOS Insecticide
AMAZE

LD50-HB	MICROGRAM/BEE	PPM		
	0.61	4.73		
		RT_{25}, RT_{40}		
FORMULATION,DOSAGE		HB	AB	LB
EC	1.0 LB AI/ACRE	>1 DAY	>1 DAY	>1 DAY

ISOFENPHOS IS HIGHLY TOXIC TO BEES.

ISOPROPYL PARATHION Insecticide

LD50-HB	MICROGRAM/BEE	PPM		
	NA	NA		
		RT_{25}, RT_{40}		
FORMULATION,DOSAGE		HB	AB	LB
EC	0.5 LB AI/ACRE	<2 HOURS	<2 HOURS	<2 HOURS
EC	1.0 LB AI/ACRE	<2 HOURS	<2 HOURS	<2 HOURS

ISOPROPYL PARATHION IS LOW IN HAZARD TO BEES.

LEAD ARSENATE Insecticide

LD50-HB	MICROGRAM/BEE	PPM		
	27.15	212.11 (ARSENICALS)		
		RT_{25}, RT_{40}		
FORMULATION,DOSAGE		HB	AB	LB
WP	3 LB AI/100 GAL	>1 DAY	>1 DAY	NA

LEAD ARSENATE CANNOT BE SAFELY APPLIED TO BLOOMING CROPS WHEN BEES ARE PRESENT.

LEPTOPHOS Insecticide
PHOSVEL

LD50-HB	MICROGRAM/BEE	PPM		
	NA	NA		
		RT_{25}, RT_{40}		
FORMULATION,DOSAGE		HB	AB	LB
EC	1.0 LB AI/ACRE	<2-3 HOURS	2-16,<2-10 HRS	>1,1 DAY

LEPTOPHOS IS MODERATELY HAZARDOUS TO HONEY BEES.

LIME-SULFUR Fungicide, Insecticide

LD50-HB	MICROGRAM/BEE	PPM		
	NA	NA		
		RT_{25}, RT_{40}		
FORMULATION,DOSAGE		HB	AB	LB
LC	3 GAL/100 GAL	<2 HOURS	NA	NA

FOR EXPLANATION OF DATA, SEE PAGES 160-161.

LIME-SULFUR IS RELATIVELY NONTOXIC TO HONEY BEES. IT ALSO REDUCES VISITATION ABOUT 50% ON THE DAY FOLLOWING APPLICATION.

LINDANE   Insecticide

LD50-HB	MICROGRAM/BEE	PPM		
	NA	NA		

		RT_{25}, RT_{40}		
FORMULATION,DOSAGE		HB	AB	LB
EC	0.9 LB AI/ACRE	>2 DAYS	NA	NA

LINDANE IS HIGHLY TOXIC TO HONEY BEES.

MALATHION   Insecticide

LD50-HB	MICROGRAM/BEE	PPM		
	0.73	5.67		

		RT_{25}, RT_{40}		
FORMULATION,DOSAGE		HB	AB	LB
EC	1.0 LB AI/ACRE	6,2 HOURS	NA	2.5,2 DAYS
D, WP	1.0 LB AI/ACRE	2,1.5 DAYS	NA	NA
95%TECH ULV 8 FL OZ/ACRE OR MORE				
		5.5,4 DAYS	NA	6.5,5 DAYS
95%TECH ULV 3 FL OZ/ACRE OR LESS				
		3,2 HOURS	NA	NA
G	1.0 LB AI/ACRE	<2 HOURS	NA	NA

MALATHION TECHNICAL GRADE AT THE GRASSHOPPER CONTROL RATE OF 8 FLUID OUNCES/ACRE RETAINS A LONG RESIDUAL HAZARD TO BEES; WHILE AT THE MOSQUITO ABATEMENT RATE OF 3 FLUID OUNCES/ACRE IT IS MINIMALLY HAZARDOUS (BOTH APPLIED ULTRA LOW VOLUME WITHOUT DILUTION IN WATER).

MALONOBEN   Insecticide, Miticide

LD50-HB	MICROGRAM/BEE	PPM		
	NA	NA		

		RT_{25}, RT_{40}		
FORMULATION,DOSAGE		HB	AB	LB
EC	1.0 LB AI/ACRE	<2 HOURS	<2 HOURS	16,11 HOURS
WP	1.0 LB AI/ACRE	<2 HOURS	<2 HOURS	<1 DAY,8 HOURS

MALONOBEN IS LOW IN HAZARD TO HONEY BEES.

MENAZON   Insecticide
SAPHICOL

LD50-HB	MICROGRAM/BEE	PPM
	NA	NA

FOR EXPLANATION OF DATA, SEE PAGES 160-161.

FORMULATION,DOSAGE	HB	RT_{25}, RT_{40} AB	LB
WP 1.0 LB AI/ACRE	<2 HOURS	NA	<2 HOURS

MENAZON IS LOW IN HAZARD TO BOTH HONEY BEES AND ALFALFA LEAFCUTTING BEES.

MEPHOSFOLAN   Insecticide, Systemic
CYTROLANE

LD50-HB	MICROGRAM/BEE	PPM
	NA	NA

FORMULATION,DOSAGE	HB	RT_{25}, RT_{40} AB	LB
G 1-3 LB AI/ACRE	<2 HOURS	NA	NA

MEPHOSFOLAN GRANULES ARE ESSENTIALLY NONTOXIC TO HONEY BEES.

METHAMIDOPHOS   Insecticide
MONITOR, HAMIDOP, TAMARON

LD50-HB	MICROGRAM/BEE	PPM
	1.37	10.7

FORMULATION,DOSAGE	HB	RT_{25}, RT_{40} AB	LB
LS,EC 0.5 LB AI/ACRE	6,4 HOURS	18,13 HOURS	2,1 DAYS
LS,EC 1.0 LB AI/ACRE	8-24,5-16 HOURS	1-5,1-3 DAYS	1-5,1->5 DAYS

METHAMIDOPHOS IS HIGHLY TOXIC TO BEES.

METHIDATHION   Insecticide
SUPRACIDE

LD50-HB	MICROGRAM/BEE	PPM
	0.24	1.85

FORMULATION,DOSAGE	HB	RT_{25}, RT_{40} AB	LB
EC 1.0 LB AI/ACRE	1-3, 0.5-3 DAYS	0.5-2.5 DAYS	0.5-3,0.5-2 DAYS

E.L. ATKINS GETS 2.5 DAYS RESIDUAL ACTION IN SOUTHERN CALIFORNIA.

METHIOCARB   Insecticide, Bird Repellent
MESUROL

LD50-HB	MICROGRAM/BEE	PPM
	0.37	2.9

FORMULATION,DOSAGE	HB	RT_{25}, RT_{40} AB	LB
WP 1.25 LB AI/100 GAL	>3 DAYS	NA	NA

METHIOCARB CANNOT BE SAFELY APPLIED TO BLOOMING CROPS WHEN BEES ARE PRESENT.

METHOMYL   Insecticide
LANNATE, NUDRIN

LD50-HB	MICROGRAM/BEE	PPM		
	1.29	10.08		

			RT_{25}, RT_{40}	
FORMULATION,DOSAGE		HB	AB	LB
SP,LS	0.45 LB AI/ACRE	2,<2 HOURS	5-8,2-4 HOURS	6-15,2-6 HOURS
SP,LS	0.90 LB AI/ACRE	6,3 HOURS	24,9 HOURS	24,9 HOURS
SP,LS	0.225 LB AI/ACRE	<2 HOURS	2,<2 HOURS	2,<2 HOURS
D	0.45 LB AI/ACRE	>1 DAY	NA	NA
D	1.0 LB AI/ACRE	>1 DAY	>1 DAY	>1 DAY

DOSAGES AS LOW AS 0.225 LB AI /ACRE HAVE CAUSED SEVERE KILLS OF LEAFCUT-TING BEES, PROBABLY BECAUSE THE BEES HAD BEEN IN THE FIELDS FOR SEVERAL WEEKS AND WERE MORE SUSCEPTIBLE THAN WHEN THEY FIRST EMERGED. METHOMYL CAN CAUSE BOTH REPELLENCY AND TOXICITY PROB-LEMS FOR HONEY BEES THROUGH NECTAR CONTAMINATION IN RED RASPBER-RIES.

METHOXYCHLOR   Insecticide
MARLATE

LD50-HB	MICROGRAM/BEE	PPM		
	NA	NA		

			RT_{25}, RT_{40}	
FORMULATION,DOSAGE		HB	AB	LB
EC	3 LB AI/ACRE	<2 HOURS	<2 HOURS	2-4,<2 HOURS
WP	3 LB AI/ACRE	<2 HOURS	NA	8,5 HOURS

METHOXYCHLOR IS RELATIVELY NONTOXIC TO BEES(EC FORMULATION).

METHYL PARATHION   Insecticide
METACIDE, BLADAN M, FOLIDOL, PENNCAP-M

LD50-HB		MICROGRAM/BEE	PPM
EC		0.11	0.86
PENNCAP-M		0.24	1.88

			RT_{25}, RT_{40}	
FORMULATION,DOSAGE		HB	AB	LB
EC	0.5 LB AI/ACRE	<1-3 DAYS	21,17 HOURS	>3 DAYS
EC	1.0 LB AI/ACRE	>3 DAYS	NA	>5 DAYS
PENNCAP-M	0.5 LB AI/ACRE	>4 DAYS	NA	>6 DAYS
PENNCAP-M	1.0 LB AI/ACRE	>7 DAYS	NA	8 DAYS

E.L. ATKINS GETS GREATER THAN 5 DAYS RESIDUAL ACTION IN SOUTHERN CALIFORNIA WITH PENNCAP-M, BUT ONLY 0.5 DAY WITH THE EC FORMULATION.

MEVINPHOS   Insecticide
PHOSDRIN

LD50-HB	MICROGRAM/BEE	PPM
	0.3	2.38

FOR EXPLANATION OF DATA, SEE PAGES 160-161.

FORMULATION,DOSAGE	HB	RT_{25}, RT_{40} AB	LB
EC 0.5 LB AI/ACRE	<5,<2 HOURS	<5,<2 HOURS	<5,<2 HOURS

E.L. ATKINS GETS LESS THAN 1 TO 1.5 DAYS RESIDUAL ACTION IN SOUTHERN CALIFORNIA.

MEXACARBATE   Insecticide
ZECTRAN

LD50-HB	MICROGRAM/BEE	PPM
	0.3	2.35

FORMULATION,DOSAGE	HB	RT_{25}, RT_{40} AB	LB
EC 0.75 LB AI/ACRE	1-2 DAYS	NA	NA

E.L. ATKINS GETS 3 DAYS RESIDUAL ACTION IN SOUTHERN CALIFORNIA.

MONOCROTOPHOS   Insecticide
AZODRIN

LD50-HB	MICROGRAM/BEE	PPM
	0.36	2.79

FORMULATION,DOSAGE	HB	RT_{25}, RT_{40} AB	LB
LS 0.4 LB AI/ACRE	>1 DAY	NA	NA
LS 0.5 LB AI/ACRE	>1 DAY	>1 DAY	>1 DAY
LS 0.6 LB AI/ACRE	>2 DAYS	NA	NA
LS 0.8 LB AI/ACRE	>2 DAYS	NA	NA

E.L. ATKINS GETS 2 TO 3.5 DAYS RESIDUAL ACTION IN SOUTHERN CALIFORNIA.

NALED   Insecticide
DIBROM

LD50-HB	MICROGRAM/BEE	PPM
	0.49	3.79

FORMULATION,DOSAGE	HB	RT_{25}, RT_{40} AB	LB
EC 1.0 LB AI/ACRE	12-20,4-15 HRS	1-2,0.2-1 DAYS	1-4.5,1-4 DAYS
EC 0.5 LB AI/ACRE	2 HOURS	>2 HOURS	12 HOURS
D OR WP 1.0 LB AI/ACRE	>1 DAY	NA	NA

NALED, AT 1.0 LB AI/ACRE CANNOT BE SAFELY APPLIED TO BLOOMING CROPS WHEN BEES ARE PRESENT.

NICOTINE SULFATE   Insecticide

LD50-HB	MICROGRAM/BEE	PPM
	NA	NA

FORMULATION,DOSAGE	HB	RT_{25}, RT_{40} AB	LB
S 0.4 PT AI/100 GAL	<2 HOURS	NA	NA

NICOTINE SULFATE IS RELATIVELY NONTOXIC TO HONEY BEES

OIL SPRAYS (SUPERIOR TYPE) Insecticide

LD50-HB	MICROGRAM/BEE	PPM
	NA	NA

FORMULATION,DOSAGE	HB	RT_{25}, RT_{40} AB	LB
LC 1.5-2 GAL/100 GAL	<3 HOURS	NA	NA

SUPERIOR OIL SPRAYS ARE ESSENTIALLY NONTOXIC TO HONEY BEES BUT DIRECT APPLICATION ON FORAGING BEES WILL CAUSE SOME KNOCKDOWN.

OMETHOATE   Insecticide, Miticide
FOLIMAT

LD50-HB	MICROGRAM/BEE	PPM
	NA	NA

FORMULATION,DOSAGE	HB	RT_{25}, RT_{40} AB	LB
EC 1.0 LB AI/A	>1 DAY	NA	NA

OMETHOATE IS HIGHLY HAZARDOUS TO HONEY BEES.

OVEX   Miticide
OVOTRAN,OVOCHLOR, SAPPIRAN

LD50-HB	MICROGRAM/BEE	PPM
	NA	NA

FORMULATION,DOSAGE	HB	RT_{25}, RT_{40} AB	LB
EC 2.0 LB AI/ACRE	<2 HOURS	NA	NA

OVEX IS ESSENTIALLY NONTOXIC TO BEES.

OXAMYL   Insecticide
VYDATE

LD50-HB	MICROGRAM/BEE	PPM
	10.26	80.16

FORMULATION,DOSAGE	HB	RT_{25}, RT_{40} AB	LB
LS 1.0 LB AI/ACRE	12,5 HOURS	2,1.5 DAYS	2 DAYS
LS 0.5 LB AI/ACRE	3,<2 HOURS	9,6 HOURS	9,4 HOURS
LS 0.25 LB AI/ACRE	<2 HOURS	<2 HOURS	<2 HOURS

OXAMYL AT 0.5 LB AI/ACRE OR LESS IS RELATIVELY LOW HAZARD TO HONEY BEES WHEN APPLIED DURING LATE EVENING, NIGHT OR EARLY MORNING.

FOR EXPLANATION OF DATA, SEE PAGES 160-161.

OXYDEMETONMETHYL   Insecticide, Miticide, Systemic
METASYSTOX-R

LD50-HB	MICROGRAM/BEE	PPM
	2.86	22.3

RT_{25}, RT_{40}

FORMULATION,DOSAGE		HB	AB	LB
EC	0.5 LB AI/ACRE	<2-5,<2 HOURS	<2 HOURS	<2-8,<2 HOURS
EC	0.375 LB AI/ACRE	<2 HOURS	<2 HOURS	<2 HOURS

OXYDEMETONMETHYL IS LOW IN HAZARD TO BEES WHEN APPLIED DURING LATE EVENING, NIGHT OR EARLY MORNING.

OXYTHIOQUINOX   Miticide, Insecticide, Fungicide
MORESTAN, FORSTAN

LD50-HB	MICROGRAM/BEE	PPM
	NA	NA

RT_{25}, RT_{40}

FORMULATION,DOSAGE		HB	AB	LB
WP	0.25 LB AI/100 GAL	<2HOURS	NA	NA
WP	0.125 LB AI/100 GAL	<2 HOURS	NA	NA

OXYTHIOQUINOX IS RELATIVELY NONTOXIC TO HONEY BEES.

PARATHION   Insecticide

LD50-HB	MICROGRAM/BEE	PPM
	0.18	1.37

RT_{25}, RT_{40}

FORMULATION,DOSAGE		HB	AB	LB
EC	0.5 LB AI/ACRE	13-18,10-13 HRS	1,0.5-1 DAYS	0.5-3,0.5-2 DAYS
D	0.5 LB AI/ACRE	>1 DAY	NA	NA

E.L. ATKINS GETS 1 DAY OF RESIDUAL ACTION IN SOUTHERN CALIFORNIA.

PERMETHRIN   Insecticide
AMBUSH, POUNCE

LD50-HB	MICROGRAM/BEE	PPM
	0.16	1.24

RT_{25}, RT_{40}

FORMULATION,DOSAGE		HB	AB	LB
EC	0.2 LB AI/ACRE	>3 DAYS	2,1.5 DAYS	>3 DAYS
EC	0.1 LB AI/ACRE	0.5-2,0.5-1.5DY	1-2,1 DAYS	0.5-3,0.5-2.5 DAYS
EC	0.05 LB AI/ACRE	18,22 HOURS	>1 DAY,8 HOURS	NA

PERMETHRIN IS SAFENED BY ITS REPELLENCY TO BEES UNDER ARID CONDITIONS. HOWEVER, IT IS REPORTED TO BE MUCH MORE HAZARDOUS IN HUMID AREAS. E.L. ATKINS GETS GREATER THAN 5 DAYS RESIDUAL ACTION IN SOUTHERN CALIFORNIA.

PHENTHOATE   Insecticide
CIDIAL, ELSAN

LD50-HB	MICROGRAM/BEE	PPM		
	NA	NA		
			RT_{25}, RT_{40}	LB
EC	1.0 LB AI/ACRE	>1 DAY	>1 DAY	>1 DAY
EC	0.5 LB AI/ACRE	>1 DAY	>1,1 DAY	>1 DAY

PHENTHOATE IS HIGHLY TOXIC TO BEES.

PHORATE   Insecticide, Miticide, Systemic
THIMET

LD50-HB	MICROGRAM/BEE	PPM		
	10.25	80.08		
			RT_{25}, RT_{40}	
FORMULATION,DOSAGE		HB	AB	LB
EC	1.0 LB AI/ACRE	5 HOURS	NA	NA
G	2.0 LB AI/ACRE	<2 HOURS	<2 HOURS	<2 HOURS

PHORATE GRANULES ARE NONTOXIC TO BEES.

PHOSALONE   Insecticide, Miticide, Systemic
ZOLONE, RUBITOX

LD50-HB	MICROGRAM/BEE	PPM		
	8.97	70.08		
			RT_{25}, RT_{40}	
FORMULATION,DOSAGE		HB	AB	LB
EC	1.5 LB AI/ACRE	<2 HOURS	<2 HOURS	<2 HOURS
EC	1.0 LB AI/ACRE	<2 HOURS	<2 HOURS	<2 HOURS

PHOSALONE IS LOW IN HAZARD TO BEES AND CAN BE APPLIED DURING LATE EVENING, NIGHT, OR EARLY MORNING.

PHOSFOLAN   Insecticide, Systemic
CYOLANE

LD50-HB	MICROGRAM/BEE	PPM		
	NA	NA		
			RT_{25}, RT_{40}	
FORMULATION,DOSAGE		HB	AB	LB
EC	1.0 LB AI/ACRE	1 DAY	NA	NA
G	1.3 LB AI/ACRE	<2 HOURS	NA	NA

PHOSFOLAN IS MODERATELY TOXIC TO HONEY BEES.

184

FOR EXPLANATION OF DATA, SEE PAGES 160-161.

PHOSMET   Insecticide, Miticide
IMIDAN, PROLATE

LD50-HB MICROGRAM/BEE PPM
                1.13   8.83

FORMULATION,DOSAGE		HB	RT$_{25}$, RT$_{40}$ AB	LB
EC	2.0 LB AI/ACRE	>3 DAYS	NA	NA
EC	1.0 LB AI/ACRE	2-3 DAYS	3-5,2.5-4.5 DAYS	3-5 DAYS
EC	0.5 LB AI/ACRE	18,8 HOURS	NA	NA
WP	0.5 LB AI/ACRE	>1 DAY	NA	NA
WP	1.0 LB AI/ACRE	>3 DAYS	NA	NA
WP	2.0 LB AI/ACRE	>3 DAYS	NA	NA

PHOSMET IS HIGHLY TOXIC TO BEES.

PHOSPHAMIDON   Insecticide, Systemic
DIMECRON

LD50-HB MICROGRAM/BEE PPM
                1.45   11.33

FORMULATION,DOSAGE		HB	RT$_{25}$, RT$_{40}$ AB	LB
EC	0.5 LB AI/ACRE	1-2,1 DAYS	1-2,1 DAYS	2,1.5 DAYS

PHOSPHAMIDON CANNOT BE SAFELY APPLIED TO BLOOMING CROPS WHEN BEES ARE PRESENT. E.L. ATKINS GETS 2-5 DAYS OF RESIDUAL ACTION IN SOUTHERN CALIFORNIA.

PHOSTEX   Insecticide

LD50-HB MICROGRAM/BEE PPM
            NA   NA

FORMULATION,DOSAGE		HB	RT$_{25}$, RT$_{40}$ AB	LB
EC	1.0 LB AI/ACRE	<2 HOURS	NA	<2 HOURS

PHOSTEX IS LOW IN HAZARD TO BEES.

PHOXIM   Insecticide
BAYTHION, VOLATON

LD50-HB MICROGRAM/BEE PPM
                NA   NA

FORMULATION,DOSAGE		HB	RT$_{25}$, RT$_{40}$ AB	LB
EC	1.5 LB AI/ACRE	1 DAY	1.5,1 DAYS	2.5,2 DAYS

PHOXIM IS HIGHLY TOXIC TO BEES.

PIRIMICARB   Insecticide, Systemic
PIRIMOR

LD50-HB   MICROGRAM/BEE   PPM
　　　　　　　NA　　　　　　　　NA

FORMULATION,DOSAGE		HB	RT_{25}, RT_{40} AB	LB
WP	1.0 LB AI/ACRE	<2 HOURS	NA	NA
WP	0.5 LB AI/ACRE	<2 HOURS	<2 HOURS	<2 HOURS
WP	0.25 LB AI/ACRE	<2 HOURS	<2 HOURS	<2 HOURS
EC	0.25 LB AI/ACRE	<2 HOURS	<2 HOURS	<2 HOURS
EC	0.5 LB AI/ACRE	<2 HOURS	<2 HOURS	<2 HOURS

PIRIMICARB IS RELATIVELY NONTOXIC TO BEES.

PIRIMIPHOS-ETHYL   Insecticide
PRIMICID

LD50-HB   MICROGRAM/BEE   PPM
　　　　　　　　NA　　　　　　　NA

FORMULATION,DOSAGE		HB	RT_{25}, RT_{40} AB	LB
EC	1.0 LB AI/ACRE	1 DAY	NA	2.5,2 DAYS

PIRIMIPHOS-ETHYL IS HIGHLY TOXIC TO BEES.

PIRIMIPHOS-METHYL   Insecticide
ACTELLIC

LD50-HB   MICROGRAM/BEE   PPM
　　　　　　　NA　　　　　　　　NA

FORMULATION,DOSAGE		HB	RT_{25}, RT_{40} AB	LB
EC	0.5 LB AI/ACRE	7,<2 HOURS	NA	9,<2 HOURS

PIRIMIPHOS-METHYL IS MODERATELY TOXIC TO BEES.

PROFENOFOS   Insecticide
CURACRON

LD50-HB   MICROGRAM/BEE   PPM
　　　　　　　3.46　　　　　　　27.03

FORMULATION,DOSAGE		HB	RT_{25}, RT_{40} AB	LB
EC	1.0 LB AI/ACRE	9,<2 HOURS	2,1 DAYS	2,1 DAYS

PROFENFOS IS MODERATELY TOXIC TO HONEY BEES, BUT HIGHLY TOXIC TO ALKALI AND ALFALFA LEAFCUTTING BEES.

FOR EXPLANATION OF DATA, SEE PAGES 160-161.

PROPARGITE   Miticide
COMITE, OMITE

LD50-HB	MICROGRAM/BEE	PPM
	NA	NA

FORMULATION,DOSAGE		RT_{25}, RT_{40}	
	HB	AB	LB
EC 1.5-2.0 LB AI/ACRE	<2 HOURS	<2 HOURS	<2 HOURS

PROPARGITE IS RELATIVELY NONTOXIC TO BEES.

PROPOXUR   Insecticide
BAYGON, BLATTANEX, SUNCIDE

LD50-HB	MICROGRAM/BEE	PPM
	1.34	10.47

FORMULATION,DOSAGE		RT_{25}, RT_{40}	
	HB	AB	LB
EC 0.75 LB AI/ACRE	16-22,11-18 HRS	1 DAY	1 DAY
WP(ULV) 0.156 LB AI/ACRE	<2 HOURS	NA	<2 HOURS
G 1.0 LB AI/ACRE	<2 HOURS	<2 HOURS	<2 HOURS

PROPOXUR APPLIED AT THE MOSQUITO ABATEMENT RATE OF 0.156 LB AI / ACRE IN 1.5 PINTS SPRAY OIL / ACRE IS MINIMALLY HAZARDOUS TO BEES.

PROPYL THIOPYROPHOSPHATE   Insecticide, Miticide
ASPON

LD50-HB	MICROGRAM/BEE	PPM
	NA	NA

FORMULATION,DOSAGE		RT_{25}, RT_{40}	
	HB	AB	LB
EC 0.5 LB AI/ACRE	<2 HOURS	NA	NA

PROPYL THIOPYROPHOSPHATE IS LOW IN HAZARD TO HONEY BEES.

PROTHIOPHOS   Insecticide
TOKUTHION, BIDERON, TOYOTHION

LD50-HB	MICROGRAM/BEE	PPM
	NA	NA

FORMULATION,DOSAGE		RT_{25}, RT_{40}	
	HB	AB	LB
EC 1.5 LB AI/ACRE	>1 DAY	NA	NA
EC 1.0 LB AI/ACRE	10,8 HOURS	NA	NA
EC 0.5 LB AI/ACRE	<2 HOURS	NA	NA

PROTHIOPHOS IS LOW TO HIGH IN HAZARD TO HONEY BEES, DEPENDING ON DOSAGE.

PYRETHRUM   Insecticide
PYRENONE, PYROCIDE

LD50-HB	MICROGRAM/BEE	PPM		
	NA	NA		
			RT_{25}, RT_{40}	
FORMULATION,DOSAGE		HB	AB	LB
S 0.04 LB AI/100 GAL		<2 HOURS	NA	NA

PYRETHRUM IS RELATIVELY NONTOXIC TO HONEY BEES.

RONNEL   Insecticide
KORLAN

LD50-HB	MICROGRAM/BEE	PPM		
	5.62	43.9		
			RT_{25}, RT_{40}	
FORMULATION,DOSAGE		HB	AB	LB
EC 1.0 LB AI/ACRE		1,0.5 DAYS	NA	3,2.5 DAYS

RONNEL IS HIGHLY TOXIC TO BEES.

ROTENONE   Insecticide

LD50-HB	MICROGRAM/BEE	PPM		
	NA	NA		
			RT_{25}, RT_{40}	
FORMULATION,DOSAGE		HB	AB	LB
S,D 0.15 LB AI/ACRE		<2 HOURS	NA	NA

ROTENONE IS RELATIVELY NONTOXIC TO HONEY BEES.

RYANIA   Insecticide, Miticide, Systemic
RYANICIDE

LD50-HB	MICROGRAM/BEE	PPM		
	NA	NA		
			RT_{25}, RT_{40}	
FORMULATION,DOSAGE		HB	AB	LB
WP 0.6 LB AI/ACRE		<2 HOURS	NA	NA

RYANIA IS RELATIVELY NONTOXIC TO HONEY BEES.

SCHRADAN   Insecticide
OMPA, PESTOX, SYTAM

LD50-HB	MICROGRAM/BEE	PPM		
	NA	NA		
			RT_{25}, RT_{40}	
FORMULATION,DOSAGE		HB	AB	LB
S 1.0 LB AI/ACRE		<2 HOURS	<2 HOURS	<2 HOURS

FOR EXPLANATION OF DATA, SEE PAGES 160-161.

SCHRADAN IS LOW IN HAZARD TO BEES.

SODIUM FLUOSILICATE   Insecticide

LD50-HB	MICROGRAM/BEE	PPM
	NA	NA

FORMULATION,DOSAGE		HB	RT_{25}, RT_{40} AB	LB
BAIT	30-75 LB BAIT/ACRE	<2 HOURS	NA	NA

SODIUM FLUOSILICATE BAITS ARE NONHAZARDOUS TO BEES.

STIROFOS   Insecticide
GARDONA, RABON

LD50-HB	MICROGRAM/BEE	PPM
	1.39	10.86

FORMULATION,DOSAGE		HB	RT_{25}, RT_{40} AB	LB
EC	1.0 LB AI/ACRE	<2 HOURS	10,<2 HOURS	1 DAY

STIROFOS IS LOW IN HAZARD TO HONEY BEES.

SULFUR   Miticide

LD50-HB	MICROGRAM/BEE	PPM
	NA	NA

FORMULATION,DOSAGE		HB	RT_{25}, RT_{40} AB	LB
D	3 LB AI/ACRE	<2 HOURS	NA	NA
WP	3 LB AI/ACRE	<2 HOURS	NA	NA

SULFUR IS LOW IN HAZARD TO BEES.

SULPROFOS   Insecticide
BOLSTAR

LD50-HB	MICROGRAM/BEE	PPM
	7.22	56.4

FORMULATION,DOSAGE		HB	RT_{25}, RT_{40} AB	LB
EC	0.5 LB AI/ACRE	<2 HOURS	NA	NA
EC	1.0 LB AI/ACRE	8,6 HOURS	NA	NA
EC	1.5 LB AI/ACRE	>1 DAY	NA	NA

SULPROFOS IS LOW TO HIGH IN HAZARD TO HONEY BEES, DEPENDING ON DOSAGE.

TDE   Insecticide
ROTHANE, DDD

LD50-HB	MICROGRAM/BEE	PPM		
	NA	NA		

			RT_{25}, RT_{40}	
FORMULATION,DOSAGE		HB	AB	LB
EC	3.0 LB AI/ACRE	2 HOURS	NA	NA
WP	1.0 LB AI/ACRE	<2 HOURS	NA	NA

TDE IS LOW IN HAZARD TO HONEY BEES.

TEMEPHOS   Insecticide
ABATE, BIOTHION

LD50-HB	MICROGRAM/BEE	PPM
	1.4	10.9

		RT_{25}, RT_{40}	
FORMULATION,DOSAGE	HB	AB	LB
EC 0.5 LB AI/ACRE	<2 HOURS	<2-15,<2 HOURS	2-40,<2-18 HOURS

TEMEPHOS IS LOW IN HAZARD TO HONEY BEES.

TEPP   Insecticide
IFOS, TETRON

LD50-HB	MICROGRAM/BEE	PPM
	0.002	0.016

		RT_{25}, RT_{40}	
FORMULATION,DOSAGE	HB	AB	LB
EC 0.5 LB AI/ACRE	<5,<2 HOURS	<5,<2 HOURS	<5,<2 HOURS

E.L. ATKINS GETS 0.5 DAY RESIDUAL ACTION IN SOUTHERN CALIFORNIA.

TETRADIFON   Miticide
TEDION

LD50-HB	MICROGRAM/BEE	PPM
	NA	NA

		RT_{25}, RT_{40}	
FORMULATION,DOSAGE	HB	AB	LB
EC 0.75 LB AI/ACRE	<2 HOURS	<2 HOURS	<3,<2 HOURS

TETRADIFON IS ESSENTIALLY NONTOXIC TO BEES.

THIODICARB   Insecticide
LARVIN

LD50-HB	MICROGRAM/BEE	PPM
	7.08	55.3

FOR EXPLANATION OF DATA, SEE PAGES 160-161.

FORMULATION,DOSAGE		HB	RT_{25}, RT_{40} AB	LB
F	1.0 LB AI/ACRE	<2 HOURS	NA	8,<2 HOURS
F	0.75 LB AI/ACRE	<2 HOURS	NA	NA
F	0.5 LB AI/ACRE	<2 HOURS	NA	<2 HOURS
WP	0.25 LB AI/ACRE	<2 HOURS	<2 HOURS	<2 HOURS
WP	0.5 LB AI/ACRE	<2 HOURS	2 HOURS	2 HOURS

THIODICARB IS LOW IN HAZARD TO HONEY BEES AND LOW TO MODERATE IN HAZARD TO ALFALFA LEAFCUTTING BEES.

TOXAPHENE   Insecticide, Miticide
STROBANE-T

LD50-HB	MICROGRAM/BEE	PPM
	NA	NA

FORMULATION,DOSAGE	HB	RT_{25}, RT_{40} AB	LB
EC 3.0 LB AI/ACRE	<2-4,<2 HOURS	10,7 HOURS	3-7,1-5 DAYS
D 3.0 LB AI/ACRE	<1 DAY	NA	NA

TOXAPHENE IS LOW TO MODERATE IN HAZARD TO HONEY BEES AND ALKALI BEES, BUT HIGHLY HAZARDOUS TO ALFALFA LEAFCUTTING BEES.

TOXAPHENE + DDT   Insecticides

LD50-HB	MICROGRAM/BEE	PPM
	NA	NA

FORMULATION,DOSAGE	HB	RT_{25}, RT_{40} AB	LB
EC 1.5+3.0 LB AI/ACRE	8-32,<2-18 HRS	13-52,10-33 HRS	1-4 DAYS

THE MIXTURE OF TOXAPHENE PLUS DDT IS MUCH MORE HAZARDOUS TO BEES THAN EITHER MATERIAL USED ALONE.

TRICHLORFON   Insecticide
DYLOX, DIPTEREX, MASOTEN, NEGUVON

LD50-HB	MICROGRAM/BEE	PPM
	NA	NA

FORMULATION,DOSAGE	HB	RT_{25}, RT_{40} AB	LB
SP,LS 1.0 LB AI/ACRE	<3-6,<2-4 HOURS	6-14,2-6 HOURS	<2-5,<2-3 HOURS

TRICHLORFON IS LESS HAZARDOUS TO THE ALFALFA LEAFCUTTING BEE THAN TO THE HONEY BEE OR THE ALKALI BEE.

17   FURTHER READING

Adey, Margaret, Penelope Walker and P.T. Walker. 1986. *Pest Control Safe for Bees*. IBRA, London. 224 p.

Anderson, L.D. and E.L. Atkins, Jr. 1968. *Pesticide usage in relation to beekeeping*. Ann. Rev. Entomology. 13: 213-238.

Atkins, E.L. 1984. Injury to honey bees by poisoning. pp. 663-695. In *The Hive and the Honey Bee*. Dadant, Hamilton, Il.

Atkins, E.L., D.K. Kellum and K.W. Atkins. 1981. *Reducing pesticide hazards to honey bees*. Univ. CA Leaflet 2883. 22 p.

Beran, F. and E. Glofke. 1959. 2. *Mitteilung: Der nachweis von bienenvergiftungen*. Pflanzenschutzberichte 22: 145-171.

Cook, A.J. 1889. *Spraying with arsenites*. Michigan Agr. Exp. Sta. Bul. 53: 3-8.

Crane, E. and P. Walker. 1983. *The impact of pest management on bees and pollination*. Tropical Development and Research Institute. London, England. 84 pp.

Erickson, E.H. Jr. and B.J. Erickson. 1984. *Guidelines for future honey bee/pesticide research*. Amer. Bee J. 124(2):131-134.

Evenius, J. 1960. *Bienezucht und pflazenschutz*. Biene and Bienenzucht: 338-346.

Herman, F.A. and W.J. Brittain. 1933. Studies in bee poisoning as a phase of orchard pollination studies. Can. Dept. Agr. Bul. 162: 158-89.

Johansen, C.A. 1966. *Digest on bee poisoning, its effects and prevention.* Bee World. 47: 9-25.

Johansen, C.A. 1977. *Pesticides and pollinators.* Ann. Rev. Entomol. 22: 177-192.

Johansen, C.A. 1979. *Honeybee poisoning by chemicals: Signs, contributing factors, current problems and prevention.* Bee World 60 (3): 109-127

Johansen, C.A., D. F. Mayer, J.D. Eves and C.W. Kious. 1983 *Pesticides and bees.* Envir. Entomol. 12:1513-1518.

Mayer, D.F., and C.A. Johansen. 1985. Pollinator protection and acephate (orthene) insecticide. Amer. Bee J. 125: 207-210.

Mayer, D.F., C.A. Johansen, C.H. Shanks and K.S. Pike. 1987. Effects of fenvalerate insecticide on pollinators. J. Entomol. Soc. Brit. Col. 84: 39-45.

Mayer, D.F., I.D. Lunden and L.H. Weinstein. 1988. Evaluation of floride levels and effects on honey bees (Apis mellifera L.) Fluoride 21(3): 113-120.

McIndoo, W.E. and G.S. Demuth. 1926. *Effects on honeybees of spraying fruit trees with arsenicals.* USDA Bul. 1364: 1-32.

Musgrave, A.J. and E.H. Salkeld. 1950. *A bibliography of honey bee toxicology.* Can. Entomologist. 82: 177-179.

Mobus, B. and L. J. Connor. 1988. *The Varroa Handbook - Biology and Control.* Wicwas Press, Cheshire,CT 52 pp.

Palmer-Jones, T. 1965. *Toxicity of pesticides to honey-bees.* New Zealand Weed & Pest Control Conf. Proc. 203-7.

Palmer-Jones, T. and I.W. Forster. 1958. *Agricultural chemicals and the beekeeping industry.* New Zealand J. Agr. 97: 298-304.

Panguard, C. 1960. *Revue biblopgraphique des intoxicantions de l'abeille.* Bull. Apicot. Inform. 3: 109-180.

Shaw, F.R. 1941. *Bee poisoning: review of more important literature.* J. Econ. Entomol. 34: 16-21.

Todd, F.E. and S.E. McGregor. 1960. *The use of honey bees in the production of crops.* Ann. Rev. Entomol. 5: 265-278.

Webster, F.M. 1896. *Spraying with arsenites vs. bees.* Ohio Agr. Expt. Sta. Bul. 68: 48-53.

Wiese, I.H. 1962. *The susceptibility of honey-bees to some insecticide spray formulations used on citrus.* S. Afr. J. Agric. Sci. 5: 557-588.

Wilson, W.T., P.E. Sonnet and A. Stoner. 1980. *Pesticides and honey bee mortality. In:.* Beekeeping in the Unites States. USDA Agr. Handbook 335. pp. 129-140.

GLOSSARY 18

ARSENICALS — pesticides containing arsenic.

BEE POISONING — the accidental killing or debilitation of bees caused by the use of insecticides.

BOTANICAL PESTICIDE — a pesticide made from plants.

BROAD SPECTRUM INSECTICIDE — an insecticide that kills many different species of insects.

BROOD — a collective term for the eggs, larvae, and pupae of bees.

CARBAMATE — a synthetic organic pesticide derived from carbamic acid.

CARRIER — the inert liquid or solid material added to an active ingredient to prepare a pesticide formulation.

CHOLINESTERASE — a chemical enzyme found in animals that helps regulate the activity of nerve impulses.

COLONY — a social community of several thousand worker bees, usually containing a queen with or without drones.

COMPATIBILE — two compounds or materials that can be mixed or blended without affecting each other's performance.

CONTAMINATED POLLEN-QUEEN SUPERSE-DURE SYNDROME — series of events caused when honey bee foragers bring pesticide-contaminated pollen back to their hive: newly-emerged bees are killed; production of royal jelly is reduced; titer of queen substance is reduced; and the queen is superseded.

CORBICULAE — pollen basket, area on hindleg of honey bee adapted for carrying pellets of pollen.

DEFOLIANT — any substance intended to cause the leaves or foliage to drop from plants.

DEGRADATION — the process by which a chemical is reduced to a less complex form.

DESICCANTS — a plant drying agent and usually the foliage is killed by contact action of the chemical.

DOSAGE — quantity of pesticide applied.

DRIFT — movement of the pesticide off the target area.

FIELD BEES — bees that collect food for the colony.

FORAGERS — bees that collect food.

FORMULATION — a mixture of an active pesticide chemical with carriers, diluents, or other materials.

FUNGICIDE — any substance intended to prevent, destroy, repel, or mitigate fungi.

GLOSSA — the central extension at the end of the labium of the bee, often called the tongue.

HAZARD — is the possibility of a chemical producing adverse effects in special circumstances.

HERBICIDE — any substance intended to prevent, destroy, repel, or mitigate weeds.

HONEY — sweet viscous fluid elaborated by bees from nectar obtained from plant nectaries.

HONEYDEW — sweet excretion from aphids and scale

insects; the excess from plant sap feeding.

HYMENOPTERA — insect order to which all bees belong.

INORGANIC PESTICIDE — those pesticides lacking carbon.

INSECTICIDE — any substance intended to prevent, destroy, repel, or mitigate insects.

LABEL — all written, printed, or graphic matter on or attached to the pesticide or the immediate container as required by law.

LD_{50} — amount of pesticide in milligram per kilogram of body weight that it takes to kill 50% of the test animals.

MITICIDE — any substance intended to prevent, destroy, repel, or mitigate mites.

NECTAR — a solution of dissolved sugars secreted by the nectaries.

NECTARIES — special cells in plants from which nectar exudes.

ORGANIC COMPOUNDS — chemicals that contain carbon.

ORGANOCHLORINE INSECTICIDE — a synthetic organic pesticide that contains chlorine, carbon, hydrogen, and other elements.

ORGANOPHOSPHATES — a synthetic organic pesticide containing carbon, hydrogen, and phosphorus.

PPM (PARTS PER MILLION) — a way to express the concentration of chemicals in foods, plants and animals. One part per million equals one pound in 500 tons.

PHEROMONE — chemical substances secreted to the outside of an animal's body that convey a message when received by another of the same species.

POLLEN — fine dust-like particles containing male reproductive cells of flowers collected and used by bees directly as food newly emerged adults and indirectly as food (royal jelly) for their young.

PROBOSCIS — a collective term for the mouthparts of a bee or other sucking insects.

PROVENTRICULUS — the honey stomach in honey bees, usually the crop in other insects.

QUEEN SUBSTANCE — pheromone produced in the queen honey bee's mandibular glands.

REGURGITATION — in honey bees, occurs when dying bees expell contents of the honey stomach.

REPELLENT — any substance that causes the target species to avoid the area.

RESIDUE — pesticide material which remains on the treated surface after spraying.

RESISTANCE — the ability of an organism to suppress or retard the injurious effects of a pesticide.

RODENTICIDE — any substance intended to prevent, destroy, repel, or mitigate rats and mice.

RT_{25} — indicates the residual degradation time required to bring bee mortality down to 25% in cage test exposures to field-weathered spray deposits.

RT_{40} — indicates the residual degradation time required to bring bee mortality down to 40% in cage test exposures to field-weathered spray deposits.

SEQUENTIAL TESTING — utilizes a step system to evaluate the hazard of a pesticide to bees.

STICKER — a material added to a pesticide to increase its adherence.

SUPERSEDURE — when a colony with an old or failing queen rears a daughter to replace her and drive the old queen out of the hive; related to low titer of queen substance available within the colony.

SYNERGISM — the joint action of 2 or more pesticides that is greater than the sum of their activity when used alone.

SYSTEMIC — a pesticide is absorbed by roots, stems or leaves of a plant and translocated internally.

TODD DEAD BEE TRAP — a device fitted to the front of the honey bee colonies to collect bees that die within the colony.

TOXIC — poisonous

TOXICITY — is the inherent property of a chemical to cause adverse biological effects.

TOXICANT — a poisonous chemical

WARDECKER WATERER — A device for providing water inside a bee hive.

19

INDEX A
Chemical Index

CHEMICAL NAMES FROM
APPENDEXES I — V.

To look up a chemical listed in the hazard-rating data, look under the chemical name, listed alphabetically below. This will show you the page that that material appears in the appendix sections I through V. Trade names are in parentheses. Different formulations (D, ULC, EC, WP, etc.) are listed separately.

To look up other chemicals, other subjects, and chemicals listed in Appendix VI, use Index B - General Index.

INDEX B   20
General Index

Index to all but Appendixes I-V.

Use this section to look up general information as well as chemicals listed in Appendix VI.

For data on the bee-hazard posed by a specific pesticide, consult Index A - Chemical Index. this includes chemicals listed in Appendixes I-V.

Bold-faced numbers show the page number where chemicals are listed in Appendix VI.

Italicized numbers show the page number where terms are defined in the Glossary, page 195.

Photo Credits

Anonymous — 94, 116
Authors — 12, 14, 22, 24,26, 30, 31,32, 33, 34 bottom, 35 top, 35 bottom, 66, 76, 80, 85, 111, 123 top, 131, 136
Dewey M. Caron — 20 top
Lawrence J. Connor, Beekeeping Education Service — 9, 20 middle, 22 middle, 23, top, 31 bottom, 34 top, 45, 62, 66, 67, 68, 71, 94, 110, 129
Ohio State University Cooperative Extension — 20 bottom, 27, 78, 89, 105, 106, 121, 123 bottom, 126
Joe Ogrodnick, Entomology, New York State Agricultual Experiment Station, Geneva, NY, 21 middle
Robin Thorpe, University of California, Davis, 87
University of California Extension Entomology — 124
Ray Williamson 18 top, 18 bottom, 21 top, 21 bottom, 22 bottom, 23 middle, 23 bottom, 25, 90, Zoecon Corporation, 64

www.ingramcontent.com/pod-product-compliance
Lightning Source LLC
Chambersburg PA
CBHW061249220326
41599CB00028B/5585